The Rule of Five

The Rule of Five

Making Climate History at the Supreme Court

Richard J. Lazarus

The Belknap Press of Harvard University Press

Cambridge, Massachusetts
London, England
2020

First printing

Library of Congress Cataloging-in-Publication Data
Names: Lazarus, Richard J., author.
Title: The rule of five : making climate history at the Supreme Court /
    Richard J. Lazarus.
Description: Cambridge, Massachusetts : The Belknap Press of Harvard
    University Press, 2020. | Includes index.
Identifiers: LCCN 2019041826 | ISBN 9780674238121 (cloth)
Subjects: LCSH: United States. Supreme Court—Decision making. |
    Environmental law—United States—Interpretation and construction.
Classification: LCC KF8748 .L34 2020 | DDC 344.7304/633—dc23
LC record available at https://lccn.loc.gov/2019041826

To JEANNIE, SAM, JESSE, and

JUSTICE JOHN PAUL STEVENS,

the author of *Massachusetts v. EPA*

Contents

The Rule of Five

Prelude

On the morning of April 2, 2007, the United States Supreme Court announced its decision in *Massachusetts v. United States Environmental Protection Agency,* the most important environmental law case ever decided by the Court. The stakes were enormous. At issue was the legal authority and responsibility of the United States government to address the most pressing global environmental problem of our time—climate change.

One person, Joe Mendelson, acting very much alone, had triggered the events that had led to this moment. Fed up with the lack of action during the presidency of Bill Clinton to address the climate issue, he had decided to do something about it, even when other environmentalists had urged him to stand down. By the time the case reached the Supreme Court, George W. Bush was president and Mendelson had been joined by dozens of lawyers; together, they made up a team that called themselves "The Carbon Dioxide Warriors." They sought to do something no environmentalist had ever done before: take a case to the Supreme Court and defeat the president of the United States. Now it was up to the nine Justices of the Supreme Court to decide.

A ruling in favor of a coalition of states, local governments, and environmental groups led by Massachusetts could mean that after years of inaction, the United States would finally take action to address a global threat with potentially catastrophic consequences for which the United States, more than any other nation, was largely responsible. A loss, by contrast, could set back or even destroy efforts across the country to bring lawsuits to force the federal government to act on the climate issue. The latter prospect was why many thoughtful environmentalists had tried to prevent the *Massachusetts* case from ever going before the Justices. They were justifiably worried they might not only lose, but lose big.

This book tells the story of *Massachusetts v. EPA*. It is an unlikely yet nonetheless hopeful and inspiring story that reveals the best of Supreme Court litigation today, while acknowledging why it is hard to make law to address climate change. The *Massachusetts* story underscores the serendipitous pathways and fascinating personalities that can lead to a historic ruling. It also opens a window into the Court itself, and the many ways in which environmental cases present particular challenges both for the Justices and for the advocates who appear before them.

1

Joe

October 20, 1999. Joe Mendelson pulled his desk drawer open and reached inside. The several-hundred-page petition he had drafted more than a year earlier was still there, as he knew it would be. For all those months, the decision to file—or not—had weighed on him. It was a classic late October autumn day in the nation's capital, in the low fifties, sunny with a sporadic light drizzle. But this was not just any day. It was the day Joe Mendelson edged himself to the point of acting.

Mendelson was a public interest attorney working for the International Center for Technology Assessment, a big name for a tiny, shoestring environmental organization no one had ever heard of. In his early thirties, married with two young children, he was still youthfully fit and could easily have passed for a decade younger. Until recently his Capitol Hill office had consisted of a small bedroom overlooking an alley in a small townhouse across from a liquor store. Calling it an office may in fact be too grand; he kept his file cabinet in the bedroom closet. A city housing inspector certainly hadn't been impressed.

A few weeks earlier Mendelson and his four colleagues had relocated to an even smaller suite of rooms in a nondescript building on the Hill after the inspector had questioned the legality of their office in a residential neighborhood. As Mendelson sat staring into his desk drawer, he knew his job lacked the status and trappings of the better-heeled national environmental public interest groups. But he also knew he lacked their bureaucratic constraints. He could decide largely on his own whether legal action was needed. And he had decided that the need for such action now was obvious. The consequences, however, were far less so.[1]

The first generation of environmental public interest lawyers, in the 1970s and 1980s, had focused on easily visible and immediately harmful industrial air and water pollution. But Mendelson, who began law school in 1988, became a public interest lawyer just when alarms were beginning to sound about the less visible, more complex, and potentially catastrophic problem of what was then called global warming. In June 1988, only a few months before his first day of classes, James E. Hansen, a National Aeronautics and Space Administration scientist, testified before Congress that the earth was warming as a result of the "greenhouse effect"—the absorption in the upper atmosphere of heat from the sun by certain human-produced gases. Hansen spoke in the measured words of a scientist, but with such ominous and dramatic import that the *New York Times* published a front-page article on his testimony and included a graph produced by Hansen and a colleague showing that the four warmest years in the past century had all occurred in the 1980s. "The greenhouse effect is here," Hansen declared.[2]

"Yesterday's Gone"

When Senator Al Gore from Tennessee was elected vice president in November 1992, environmentalists had reason to be pleased. Gore had written "the book" on climate change—*Earth in the Balance*—

proposing an ambitious, if not radical, "Global Marshall Plan" to avert environmental devastation.[3] In his book Gore did not mince words on the urgency of the problem and the need for immediate and sweeping governmental efforts. "We must take bold and unequivocal action," he insisted. The "rescue of the environment" must be no less than "the central organizing principle for civilization." "Complacency" can no longer be afforded, Gore warned, because the nation, indeed the entire world, is "facing a rapidly deteriorating environment" that "threatens absolute disaster."[4]

What was more, the environmentalists knew that Clinton had chosen Gore to be his running mate not despite Gore's boldly stated environmental ambitions but because of them. Clinton picked Gore in large part because Gore's book bolstered his chances of winning the presidency. He knew that his own environmental record as governor of Arkansas was anything but clean. Indeed, it bordered on the abysmal, which is why founding members of the Sierra Club's Arkansas chapter resigned in protest when the national Sierra Club organization endorsed Clinton for president.[5]

Governor Bill Clinton's administration had burned toxic waste at Superfund hazardous waste sites, failed to crack down on massive water contamination caused by the state's powerful hog and poultry industries, declined to prevent clear-cutting of forests, and proposed a water quality standard for the toxic chemical dioxin that was one hundred times less restrictive than the standard recommended by United States Environmental Protection Agency (EPA).[6] Clinton proudly touted himself as the "jobs" candidate—a central part of his celebrated campaign theme, "It's the economy, stupid." Clinton was a self-described "new Democrat" who could win a national election. He frequently drew contrasts between himself and the left-leaning Democratic candidates who had lost the three previous elections to Ronald Reagan (twice) and George H. W. Bush.[7]

But Clinton was well aware that his "new Democrat" status created vulnerabilities within the more liberal base of the party, whose votes and enthusiasm he would need in November. That is where Gore fit in. Clinton hoped Gore could help him with those liberal Democrats who were naturally wary of a southern, moderate governor from Arkansas. When Gore's presence on the ticket generated mocking criticism from his counterpart on the Republican ticket, Senator Dan Quayle from Indiana—who described *Earth in the Balance*'s call for sweeping legal and policy changes to address climate change as "bizarre," "detached from reality," and "devoid of common sense"[8]—political commentators were divided on whether Clinton's choice would prove to be a "brilliant stroke" or "stupid."[9]

Before the book's publication, Gore's own political future was unclear, given his dismal showing when he had run for president in 1988. Gore had known, during his short-lived campaign, that he had little chance of winning. He was only forty years old, three years into his first term as senator from Tennessee. After eight years of the Reagan presidency, the Democratic field of presidential candidates was crowded with more senior party leaders. Gore's aim was to make a good enough showing to set himself up for a more effective run one or two election cycles later.

But his plan backfired. That he had presented himself as presidential at such a young age was seen as raw political hubris, suggesting that he lacked both maturity and seriousness. He emerged from the process a failed candidate rather than a promising future one. He undermined his claim to be a Democrat capable of winning southern states by losing to the civil rights leader Jesse Jackson in several Democratic presidential primaries in the South, and he did little better in the Northeast. After garnering only 10 percent of the vote in the New York primary, Gore dropped out. Chastened by his poor showing in 1988, and no longer perceived as a viable national candidate for the White House, Gore did not even campaign for the presidency in 1992.[10]

Instead, he wrote and published *Earth in the Balance* to accomplish a political makeover. And it worked. The book was on the

New York Times Best Sellers list for much of 1992 and effectively rebooted Gore's political career. The book's luster with the liberal wing of the Democratic Party was sufficient to overcome those who strongly advised Clinton against a running mate from a nearby southern state who, like Clinton himself, was youthful and lacked gravitas. The Clinton-Gore campaign embraced its youthfulness, promising a generational change, underscored by the campaign's official song, Fleetwood Mac's "Don't Stop," and its refrain: "Yesterday's gone, yesterday's gone."

Clinton and Gore's success in November 1992 did more than place Gore as the president's partner in the White House. The election catapulted him into the pole position for the presidency in the 2000 election. Ironically, those very same presidential ambitions simultaneously doomed Gore's willingness to take on the climate issue during his eight years as vice president with anything close to the vigor he had promised. In his book, Gore had presciently described his own limitations: "I have become very impatient with my own tendency to put a finger to the political winds and proceed cautiously," he wrote. He promised that the compelling nature of the climate problem would make him act differently this time: "The integrity of the environment is not just another issue to be used in the political games for popularity, votes or attention," he wrote. "The time has long since come to take more political risks—and endure much more political criticism—by proposing tougher, more effective solutions and fighting hard for their enactment."[11]

Gore championed many important environmental initiatives while in office—but not on climate change, the issue he had described as the most pressing of all. On the climate issue, Gore, in significant respects, did the exact opposite of what he had preached. He declined to take the "political risk,"[12] deciding instead to maintain a low profile in order to enhance his own prospects for the presidency in 2000.

Gore's retreat from climate change advocacy in favor of electability was evident as early as the fall of 1992 at the Democratic

National Convention, when he sensed that the strong views expressed in his book might hurt him politically. His own campaign manager even sought to distance him from the book's promises by stating that the book had merely "laid th[em] out as a series of ideas to be thought about and contemplated" rather than as formal proposals for action.[13] As vice president, Gore similarly declined to embrace the role of zealous advocate for climate change regulation within the administration, in the face of political opposition. To be sure, he persuaded Clinton to appoint to significant environmental positions within the administration former congressional staffers who shared his environmental concerns, including Carol Browner as administrator of the EPA and Katie McGinty as chair of the White House Council on Environmental Quality. But when confronted by Republicans in Congress and climate skeptics within the Clinton administration, Gore and his team mostly allowed themselves to be pushed to the side.

Mendelson and other environmentalists were unforgiving. By December, Gore found himself the target of barbed criticism by the same environmentalists who had championed his candidacy five years earlier. They viewed the United States' position on greenhouse gas emissions reduction as little different from what it had been under President Bush. Gore's balancing act was further undermined when the two State Department officials who were leading the international climate negotiations on behalf of the United States (Assistant Secretary of State Eileen Claussen and Undersecretary of State Tim Wirth) both abruptly left the team, reportedly out of frustration with both Gore and Clinton, shortly before the United Nations summit for international climate negotiations scheduled for Kyoto, Japan, in 1997.[14]

At the last minute, President Clinton sent Gore to present the U.S. position. Gore's charge in Kyoto was the opposite of the bold, affirmative action that his book had argued was essential. It amounted to a retreat. The White House tasked him with dissuading other nations from agreeing to emissions reductions that would be too stringent to be politically palatable back in the United States. The European Union's proposal called for significantly greater emissions reductions than those Gore was touting on behalf of the United States.[15]

The vice president found himself pushed and pulled by two conflicting political constituencies: the short-term economic concerns of the labor movement, whose support he would need for the 2000 national elections, and the environmental concerns of the progressive Democratic base, whose support would likely play an outsized role in the Democratic nomination process. Gore was himself especially concerned about labor's vocal opposition to tougher climate controls. The United Mine Workers union funded a study that produced the exaggerated claim that the climate change agreement being debated in Kyoto would cost 1.6 million jobs in the United States. As a presidential hopeful, Gore could ill afford to create an enemy of the labor movement.[16]

Because of the insistence of the United States negotiating team in Kyoto, the other nations ultimately agreed to relax the amount of overall emissions reductions required by individual nations. In the immediate aftermath of the signing of the Kyoto Protocol, the *Washington Post* reported that any public support environmentalists tried to offer Gore was belied by "their bitter disappointment and sense of betrayal. . . . [T]here was a consensus that what Gore wanted was an agreement that would cover his political bases, and not one that would protect the global climate."[17]

Gore's concerns were not illusory. Even the watered-down version of the Kyoto Protocol agreed to by the Clinton administration incurred the wrath of Congress. Before the final negotiations were concluded, the U.S. Senate voted 95–0 in favor of the "Byrd-Hagel Resolution"—named after Democratic Senator Robert Byrd of West Virginia and Republican Senator Chuck Hagel of Nebraska—which declared that the United States should not be a signatory to any protocol that did not mandate new specific commitments to reduce greenhouse gas emissions from China, Brazil, Mexico, India, and South Korea—which the pending negotiations would not do.[18]

Between 1992, when Gore published his book, and 2000, when he again ran for president, the climate problem was not standing still. Absent any meaningful program to limit greenhouse gas emissions, the economic boom of the 1990s resulted in ever-higher loadings of greenhouse gases in the earth's atmosphere. A concentration of 354 parts per million in 1990 had increased to 370 parts per million by 2000, significantly crossing the 350 parts per million threshold that scientists believed was environmentally sustainable.[19] Climate scientists across the globe were also growing ever more certain in their assessments of the dire consequences of such increased atmospheric concentrations of greenhouse gases. For the first time, scientists were able to describe how the regional impacts of climate change would vary, with dramatic increases in precipitation in some places and extreme drought in others. They detailed the severe ongoing reductions in the thickness in Arctic sea ice, the massive permanent loss of coral reefs, and the potentially devastating reduction in places where existing plant and animal species could survive. Scientists were also able to conclude with even more confidence that these potentially catastrophic climate consequences were the result of increased greenhouse gas emissions from human activities, and not due merely to natural forces.[20]

"The Queen of Clean Air"

Mendelson and other environmentalists looked to Carol Browner, Clinton's EPA administrator, to take action to restrict greenhouse gas emissions in response to the mounting scientific evidence of the serious threats posed by climate change. Gore's own protégé, only thirty-six years old when she took the helm at the EPA in January 1993, Browner was the youngest administrator in the agency's history. Appearing in striking photo-shoots in *Vogue* magazine, with her stylish, short-cropped brown hair and no-nonsense demeanor, Browner brought a fresh face and youthful energy to the agency. She was a captivating, charismatic, and effective spokesperson for environmental protection and was only the second woman to head the EPA.[21]

Browner was dogged and unrelenting. She invigorated the Clean Water Act's long-dormant water quality standards program, after decades of agency inaction. She launched Clean Air Act enforcement actions coast to coast, taking on the nation's aging and heavily polluting power plants, which for decades had used loopholes (supported by the EPA) that had allowed them to avoid either retiring or upgrading their pollution control equipment. And in 1996 she shepherded through Congress two significant environmental laws—the Food Quality Protection Act and the reauthorization of the Safe Drinking Water Act—overcoming a Congress otherwise wholly paralyzed by partisan divide.[22]

Nor did Browner shy away from confronting the president, his economic team, or congressional skeptics in her overriding desire to improve water and air quality. She was known as a fighter with a strong inside game, quite capable of outmaneuvering those who opposed her efforts. As the EPA administrator, Browner succeeded in significantly strengthening national air quality standards and related pollution control requirements. In what the *New York Times* described at the time as "a remarkable piece of bureaucratic bravura," she surmounted objections from the White House, especially the president's economic team, and issued tougher emissions rules designed to reduce unhealthy levels of ozone and particulate matter in the air. Browner's more demanding air pollution standards resulted in banner front-page headlines in the *New York Times* and the *Washington Post*. Her picture appeared in the coveted spot above the fold in the *Times,* and *Time* magazine dubbed her "the queen of clean air."[23]

Browner took no victory laps. Instead, she sprinted to the finish line. In her final weeks at the EPA, she overcame industry objections to create, for the first time ever, the legal basis for strict regulation of mercury from coal-fired power plants.[24] By the close of the Clinton administration at noon on January 20, 2001, Browner was only three days shy of her eighth anniversary at the EPA. Now forty-four years old, the EPA's youngest administrator had become its longest-serving administrator ever.

For much of her tenure, Browner faced fierce congressional opposition. In the 1994 midterm elections, the Republican Party's campaign platform—expressed by its "Contract with America"—firmly placed a bull's-eye on Browner and the EPA. The vast majority of the Contract's provisions singled out the EPA for criticism. Republican candidates who supported the Contract claimed that excessive EPA regulations were undermining economic growth and prosperity and eroding private property rights.

When the Republicans captured a majority in the House of Representatives that November, the newly elected speaker, Newt Gingrich, left no doubt that Browner's EPA was Public Enemy No. 1. Gingrich immediately targeted the EPA for massive budgetary cutbacks, and his new committee chairs compelled Browner to appear at oversight hearing after oversight hearing at which she and her agency were loudly denounced in nationally televised broadcasts.[25]

Browner's greatest congressional nemesis was the House majority whip, Tom DeLay, who, for all practical purposes, was a member of Congress *because* of his deep antipathy for the EPA. Immediately after graduating from college in 1970 with a bachelor's degree in biology, DeLay had joined the pesticide manufacturing business. Eventually he had purchased his own company, where he earned the moniker "The Exterminator." DeLay vehemently opposed the EPA pesticide regulations because of their adverse economic impact on his business profits. When first elected to Congress from his home district in southwest Texas in 1984, he quickly established his national reputation as the EPA's biggest (and loudest) critic.[26]

Nor did his harsh condemnations ebb as he rose through the ranks and became the House majority whip. Light years in political differences separated the White House and House majority whip during President Clinton's second term, as DeLay championed the House effort to impeach the president. Yet notwithstanding those differences, some common ground remained between the two camps on the question of whether the federal government should exercise authority to restrict greenhouse gas emissions to address climate change absent federal legislation. DeLay was vehemently opposed to any such regulation, and the White House, determined not to trigger a battle with Congress on the climate issue in the lead-up to the 2000 presidential election, was no less committed to reassuring DeLay that no such regulations were in the offing.[27]

On March 11, 1998, Carol Browner outmaneuvered both DeLay and her own boss. In unscripted congressional testimony that had not been approved by the White House, she staked out the historic position that the EPA already possessed statutory authority to regulate greenhouse gas emissions. The setting was a congressional appropriations subcommittee hearing to consider the EPA's budget for the upcoming fiscal year. It was not designed to be a high-profile event, despite the fact that DeLay was a member of the House appropriations subcommittee. The hearing took place in small, cramped quarters with little physical separation between the witnesses and their congressional inquisitors. As Browner later recalled, she sat so close to the committee members that it was as if they were "sitting on top of each other."[28]

DeLay's questions were unqualifiedly hostile. For her part, Browner did not shy away from being confrontational right back, easily matching DeLay's sarcasm with her own. Their exchange, to observers in the hearing room that day, resembled a classic D.C. political game of cat and mouse. The complication was that DeLay and Browner both saw themselves as the cat and the other as the mouse.

For two consecutive days DeLay used the appropriations hearing to try to pin Browner down. He repeatedly accused Browner of undertaking a clandestine effort to implement the Kyoto Protocol that the Senate had roundly denounced. DeLay trashed her actions on climate as unconstitutional. With obvious sarcasm, he demanded to know whether Browner thought it was "morally irresponsible to let constitutional scruples stand in the way of saving the world from a climate catastrophe." Browner shot back, asking if DeLay was questioning her "commitment to the Constitution." In response, DeLay threw a pocket-sized version of the Constitution at Browner.[29]

On the second day of the hearing, DeLay played what he clearly believed was his trump card; he brandished an internal EPA memo prepared by a midlevel political appointee in the EPA's Air Office that had been leaked to the press, which took the position that the EPA already possessed the legal authority under the Clean Air Act to regulate greenhouse gases. The memo had a cautionary footer on every page declaring "Predecisional, do not quote or cite," and neither the EPA's Office of General Counsel, nor Browner, nor anyone in their offices had ever heard of the memo, let alone reviewed it, when it appeared in the press. The memo nonetheless became quick fodder for conservative commentators, who denounced both its conclusions and the EPA. For DeLay, the leaked memo was the smoking gun for which he had long been looking. It supported his accusation that Browner's EPA was secretly implementing the Kyoto Protocol in defiance of Congress.[30]

Browner responded by disclaiming any knowledge of the memo's existence prior to the recent media reports. She testified that she did not even know who at the EPA had written the memo. But to the surprise of her staff, and especially her lawyers back at headquarters, she did not back away from the memo's conclusion that the Environmental Protection Agency possessed authority over greenhouse gases because they constituted Clean Air Act "air pollutants." Instead, she willingly embraced the proposition.[31]

The most telling moment of the exchange took place in response to DeLay's final question. Stung by Browner's answers, DeLay demanded that Browner supply him with a "legal opinion" to back up her position. Browner's one word response—"Certainly"—masked the significance of what had just happened.[32]

The White House, which has to approve congressional testimony by agency heads, would never have approved Browner's testifying that the EPA possessed the authority to regulate greenhouse gases. Nor would the White House ever have approved the EPA's issuing agency guidance or a regulation that took that legal position. But because of Browner's unscripted and unapproved testimony in a congressional hearing, a member of Congress had formally asked the EPA to provide Congress with a legal opinion on the question. Now, the agency had no choice but to answer. Normal bureaucratic hurdles to the EPA's issuing such an opinion, including prior White House approval, had been effectively circumnavigated.

The White House was caught off-guard by Browner's testimony, and so was her staff. Back at the EPA there was definitely a sense of "Oh my, this is great, but can we defend it?"[33] Immediately after Browner's declaration, the EPA's chief lawyer and his staff quickly went to work and in short order produced a legal opinion backing up Browner's testimony. Browner submitted the opinion to Congress and it became an official part of the congressional record.[34]

Blindsided by the unscripted action of the EPA chief, some officials in the White House and several in the president's cabinet were more than a little displeased. They demanded that Browner assure congressional Republicans (and Democrats from coal states) and industry that the EPA would not take any steps to exercise its Clean Air Act authority to restrict greenhouse gas emissions.[35]

With her hand forced, Browner made that political commitment to Congress.[36] But in his small D.C. office, Joe Mendelson had not.

2

Rocking the Boat

As the scientific evidence of the threat of climate change became increasingly clear and foreboding, Mendelson was frustrated that the Clinton administration did so little to restrict greenhouse gas emissions from its two greatest sources in the United States—motor vehicles and coal-fired power plants. The planet was on a collision course with ecological catastrophe, and the most vulnerable, poorest communities across the globe, which had played no role in creating the problem, would be the front line of casualties: famine, drought, rising sea levels, increasingly violent storms, and the destruction of low-lying islands and coastal areas.[1]

The rising cost of delay in addressing the issue was becoming clearer. Acting sooner rather than later could save human lives—as well as possibly entire species of animals and plants—and potentially trillions of dollars of property losses. Also, the longer one waited to act, the greater the accumulation of greenhouse gases in the atmosphere, and the harder it would be in the future to reduce the resulting higher atmospheric concentrations. In the United States, significant portions of major coastal cities would be submerged, beginning with New York City, Boston, San Francisco, and Miami and extending irreversibly to many others cities, such as New Orleans, Charleston, Norfolk, Galveston, Savannah, and Atlantic City.[2]

Mendelson was frustrated by the lack of more significant action by Gore, whom he and other environmentalists had enthusiastically supported in 1992. Yes, speeches by Clinton appointees expressed rising concern about climate change, and the Department of Energy had established important federal energy appliance efficiency standards that would reduce demand for electricity. But the administration's failure to impose limitations on greenhouse gas emissions by motor vehicles and power plants was inexcusable.[3]

In some respects the Clinton folks were more culpable in their neglect than the George H. W. Bush administration had been. They knew climate change was a serious problem and had acknowledged, through Browner's 1998 congressional testimony, that they had legal authority to do something about it. Yet even Browner had declined to exercise that authority. Worried about upsetting powerful Republicans in Congress or doing anything controversial on the climate issue that might hurt Gore's prospects for the presidency in 2000, the Clinton administration refused to restrict domestic greenhouse gas emissions.[4]

Mendelson was tired of hearing the EPA's same old excuses for not acting. What was the point, he thought, of having the authority to address a serious problem like climate change if one lacked the political courage to use it?[5]

The "Kitchen Sink"

Mendelson had gone to law school because he had hoped it would teach him how to be an effective advocate for social change. He'd worked for Greenpeace the summer after his first year at George Washington University. Following his graduation in 1991, he had worked for Jeremy Rifkin—a prominent, iconoclastic environmental activist who at that time was working principally on the dangers of genetically modified organisms. Mendelson worked with Rifkin until 1995, when a co-worker, Andrew Kimbrell, persuaded him to join the International Center for Technology Assessment (ICTA), a brand new organization Kimbrell had created six months earlier to fill the gap created by Congress's elimination of the federal Office of Technology Assessment—an office that had provided Congress with objective assessments of the important technical issues of the late twentieth century. The ICTA's mission was accordingly extremely broad—to provide the public with a full understanding of the impact of technology on society—and its few employees, like Joe, had great discretion to decide how to translate that mission into action.[6]

The ICTA operated on an extremely "modest" budget. It was the epitome of the small fry, a far cry from national "name brand" environmental organizations like the Natural Resources Defense Council, Environmental Defense Fund, and Sierra Club Legal Defense Fund (now Earthjustice), all of which had staffs in the hundreds, including lawyers, economists, and scientists with advanced degrees from the nation's most elite academic institutions, and annual budgets in the hundreds of millions of dollars. The ICTA's funding was orders of magnitude less, and its existence was mainly the result of the generosity of one person, an old friend of Mendelson's and Kimbrell's who shared their concern about climate change and was particularly interested in promoting cleaner, lower-emission motor vehicle technology.[7]

By 1998 Mendelson was seven years out of law school and impatient to make his mark. His frustration with the Clinton administration was matched by his disappointment with the reluctance of the national environmental organizations to challenge the Clinton EPA on the climate issue. Don't rock the boat, they cautioned. Better to wait for greener pastures under a Gore administration.[8]

But those groups were not his bosses. Mendelson was reminded of that every day he went to work. Unlike the hundreds of employees of the Natural Resources Defense Council and the Sierra

Club in offices across the country, Mendelson was one of only five employees, some of whom were not even full-time. At best, they worked from paycheck to paycheck. Sometimes there were no paychecks at all. One benefit of a shoestring budget, however, was independence to act.[9]

In late September, Mendelson began to contemplate defying the national environmental groups by confronting the Clinton administration on its inaction. On his own, he started to draft a formal petition demanding that the EPA use its existing authority under the Clean Air Act to address climate change by regulating greenhouse gas emissions from new motor vehicles. Cars and trucks, along with power plants, were responsible for the vast majority of U.S. greenhouse gas emissions, which then far exceeded those of any other country in the world.[10] Motor vehicles were Mendelson's natural target because the ICTA was then working on motor vehicle pollution standards and was not at all focused on power plants.[11]

Mendelson worked on the petition both in his office and late into the night at home, typically after reading bedtime stories to his two-year-old daughter, Anna, who shared her bedroom with her two-month-old baby sister, Quincy. In a gliding rocking chair in their newly painted pink bedroom, he drafted and redrafted in the dimmed lights as his children fell asleep. He watched Anna in her new big-kid bed only a few inches from the floor (Quincy had taken the bleached maple crib where she'd slept not long before), her sandy brown hair falling over the Raggedy Ann doll made by her great grandmother, blanket rising and falling with each breath. The sight of his two daughters quietly sleeping as he worked provided a ready reminder of why he cared so deeply about safeguarding the future.[12]

As Mendelson worked on his draft petition, he became increasingly convinced that his legal argument was straightforward and, on its face, bordered on being a slam dunk. Mendelson's petition relied on Section 202 of the Clean Air Act—a federal law passed by Congress and signed into law by President Richard Nixon in 1970—which requires the EPA administrator to regulate "emissions of any air pollutant from . . . new motor vehicles . . . which in his judgment cause, or contribute to, air pollution which may be reasonably anticipated to endanger public health or welfare."[13] Based on that provision, Mendelson's petition had to make three legal arguments, all of which he thought were essentially irrefutable.

First, he had to demonstrate that carbon dioxide, a greenhouse gas that motor vehicles emit in vast quantities, constituted an "air pollutant" within the meaning of the Clean Air Act. On this issue, Mendelson did not believe there was any room for debate. Not only did the federal statute expressly define "air pollutant" to include any chemical substance emitted into the ambient air—carbon dioxide was clearly a chemical substance emitted into the ambient air by motor vehicles—but, because of Carol Browner's congressional testimony, the EPA had already conceded that carbon dioxide fell within the Clean Air Act's definition of an "air pollutant."[14]

The second issue he would have to address in the petition was whether carbon dioxide emissions from new motor vehicles "cause or contribute" to climate change. Unlike the first, which was purely a question of law, this was purely factual. What was the impact of auto emissions on climate change?[15] Although he could not point to a prior EPA scientific finding—because the EPA had deliberately avoided ever addressing the issue—Mendelson was no less confident of his argument. Cars and trucks were two of the nation's biggest sources of carbon dioxide. And according to the recent reports by the United Nations Intergovernmental Panel on Climate Change, which themselves reflected the emerging scientific consensus, high volumes of carbon dioxide emissions were contributing to persistent high concentrations of carbon dioxide and other greenhouse gases in the atmosphere, which were causing climate change.[16]

The third and final issue Mendelson had to address in the petition was trickier, but he thought his arguments were no less strong. He had to establish, as required by the plain words of Section 202 of the Clean Air Act, that climate change may "reasonably be anticipated to endanger public health or welfare." This, too, he believed, was not hard to prove. The many United Nations scientific studies left no doubt on that score. Rising sea levels, higher temperatures, decreased drinking water supplies, and the spread of infectious diseases—all of which had been identified as effects of climate change— clearly threatened public health and welfare.[17] Mendelson didn't have to prove that climate change and these harmful effects were absolutely certain to occur. Congress deliberately used precautionary language in the Clean Air Act to guard against significant risks even if they were not certain to occur. Federal regulation was mandated so long as such adverse consequences "may" reasonably be "anticipated."[18]

Mendelson thought the law strongly favored his argument that the EPA was required to regulate carbon dioxide emissions from new cars and trucks, but he knew he had what lawyers characterize as a "procedural" problem. The argument that the EPA should be regulating greenhouse gas emissions from new cars and trucks was straightforward, but it was far less clear that any party had the right either to formally petition the EPA to regulate or, should the EPA deny the petition, to bring a lawsuit in court in order to compel the EPA to do so.

The Clean Air Act, like other federal environmental laws, is full of specific provisions that expressly provide "any person" with the right to petition the EPA administrator to regulate and take other actions upon a showing of endangerment to public health and welfare. Quite often those provisions further impose a deadline on the administrator to respond to the petition and allow for judicial review to challenge the administrator's failure to act within that deadline or deny the petition.[19] But the new motor vehicle provision of the Clean Air Act upon which Mendelson was relying (Section 202) includes no such language. It never mentions the possibility of petitioning the EPA to issue regulations limiting motor vehicle emissions.

Nor is there anywhere in the language of Section 202 any suggestion that the administrator has any obligation within any particular time frame to make that necessary threshold "judgment" or that anyone can petition the administrator to require her to do so.

In short, Mendelson did not have an obvious legal right to force the EPA to decide whether to regulate greenhouse gas emissions from new motor vehicles. His response? He embraced the "kitchen sink" approach and listed in the petition an armload of vague legal grounds for filing a petition, none of which was clearly on point. Mendelson's petition cited no less than "the Right to Petition Government Clause in the First Amendment" followed by generic, deliberately nonspecific references to a whole host of federal laws.[20] None clearly governed. The most the petition could muster in support of the First Amendment gambit was to cite a footnote to an 1875 Supreme Court ruling for the proposition that the right to petition is "logically implicit in, and fundamental to, the very idea of a republican form of government."[21] Wholly missing from the petition was any notion of a clear legal obligation of the government to respond to his petition, let alone to do so (as the petition further claimed without citation of any legal authority) promptly and in no more than 180 days.[22] Stripped of its unsupported exhortations about the EPA's duty to respond to the petition, Mendelson's gambit might fairly be characterized as a bluff.

By mid-October, Mendelson was done with the draft. Now, what should he do with it? The petition was thirty-five pages long, with 140 footnotes. It included several hundred more pages of attached exhibits of official government reports on the science of climate change in support of his arguments that greenhouse gases from motor vehicles endangered public health. Should he file the petition? What kind of blowback would he and his organization face?

Mendelson decided he would file but not right away—and he placed his draft petition in a desk drawer. His plan was to wait until the end of the Clinton presidency in order to have it appear "front and center for the next president," whom he assumed would be Al Gore.[23]

Pulling the Trigger

On Wednesday, October 20, 1999, a few weeks shy of a year before the 2000 presidential election, Mendelson decided it was time to act. Since he had put his petition in his desk drawer the previous October, the drumbeat of climate news had only gotten worse. The World Meteorological Organization had reported in December that the 1990s had managed to be even hotter than the 1980s: seven of the ten warmest years on record had occurred since 1990.[24] Rubbing salt in the wound, two weeks earlier the EPA's general counsel had testified before Congress, once again bending over backward to promise Republican leaders that the agency would not exercise its authority to regulate greenhouse gas emissions.[25] The EPA's unwillingness to take strong steps to limit greenhouse gases and the administration's incessant weaseling were too much for Mendelson. He wanted to be sure the petition was waiting for the next administration.

When Mendelson had canvassed other environmental activists, he hadn't gotten much support. Most of the national environmental groups opposed his filing and did so in harsh terms. They did not dispute the actual merits of his legal and policy arguments. They objected to his filing strictly on strategic grounds. Their widely shared fear was that any such petition could backfire politically and that it was likely to cause far more harm than good.

What environmentalists who cared about the climate should do, they argued, was not make trouble for the Clinton administration by raising the profile of climate change ahead of the election. The better strategy was to bide their time and keep anything potentially inflammatory out of the news, to make it more likely that Al Gore would be elected president in 2000. Once the election was behind him, President

Gore could effectively champion the climate issue as he had long promised to do. Yes, Vice President Gore had disappointed them, but he was a far more promising candidate than anyone the Republicans would put forward. So, in the meantime, environmentalists needed to lie low on climate and not make a lot of noise.

Mendelson, accustomed to being the outsider in these debates, was undeterred. He liked to think that his organization was into *advocacy*, not *acquiescing*. He saw it as his mission to push the limits, which included pushing the EPA to use its authority to limit greenhouse gases.[26]

Mendelson decided it was time to pull the trigger. He wanted to get the clock ticking on being able to bring a lawsuit against the EPA to force the agency to address the climate issue. And so on Wednesday, October 20, he took his draft petition out of his desk drawer, signed the signature page, and hand-stamped the front page with the date.[27]

He walked the two miles from his office that morning to deliver the petition by hand to the EPA. It was a crisp sunlit day—the kind of late autumn day when it felt especially uplifting to be outside. For good measure, he mailed a copy to Vice President Gore at the White House. It was, Mendelson later recalled, just an "ordinary day," though he filed the petition in person because of his nagging sense that his was not an ordinary filing.[28] When he handed his petition to an EPA employee that October morning, Mendelson had done what no one had ever done before: he had set the stage for a formal lawsuit to compel the United States to stop adding trillions of pounds of greenhouse gases each year to the atmosphere. He was equally motivated by thoughts about the world he would be leaving to his daughters and his desire to avert a global environmental catastrophe.

It took more than two weeks for Mendelson's petition to wind its way through the hallways of the EPA's national headquarters and reach the desk of Janie Poole, who worked in the Correspondence Office, a bland, windowless room with off-white walls, bright-white office partitions, long fluorescent lights on the ceiling, and bone-white metal filing cabinets where some half-dozen docket clerks sorted and redirected the hundreds of items of mail that arrived each day. A veteran docket clerk, Poole came across Mendelson's petition at 1:35 p.m. on November 5. She signed the petition's cover with her own name, formally stamped the document as "received," and placed it in a metal basket in a metal cart to be wheeled to the Office of General Counsel.[29]

The EPA's headquarters, built in the aftermath of the Great Depression, had decades earlier been occupied by the Post Office Department. The cart's pathway from Janie Poole's desk to the Office of General Counsel ran past a colorful Works Progress Administration mural that contrasted sharply with the otherwise monotonous offices and hallways. The mural, by Alfredo Di Giorgio Crimi, depicted in ancient fresco style postal workers adopting "new technologies and modes of transportation" in order "to meet the demands of 1930s America, with its sprawling suburbs and distant territories." It showed a grim-faced postman handing a letter to a (no less grim-faced) young girl—surrounded by various transportation technologies, including a train, horse-drawn truck, bicycle, and hand-truck dolly. Also included in the mural was a metal delivery cart that could easily have been mistaken for the cart from the EPA's Correspondence Office that, almost seventy years later, was now delivering Mendelson's stamped petition to its next destination.[30]

The EPA Office of General Counsel proved to be only the first of many stops. Seven years after it was diligently processed by Ms. Poole, Mendelson's signed and stamped petition would reach the United States Supreme Court. But it had a long way to go before then, and there was little reason, on that crisp autumn day, to expect that anyone would pay it much attention.

3

A Turd

Joe Mendelson's petition landed at the EPA in the fall of 1999 with a thud. The wounds of the impeachment trial were still visible throughout D.C. eight months after President Clinton's acquittal by the Senate, and the coming election was already casting a long shadow, with Texas governor George W. Bush and Vice President Gore the frontrunners for their respective political parties. One EPA political appointee stated at the time: "What do they expect us to do? We can't grant this petition given these politics."[1]

No one at the EPA was surprised that the petition had been filed by a peripheral organization they had never heard of. The general consensus, as one insider put it, was that "the big boys," meaning the national environmental groups with more resources, "wouldn't do it."[2] The mainstream environmental groups would know that asking the EPA to take action on the climate issue at that time "was a mistake."[3]

So the Clinton EPA decided to ignore the petition. One EPA appointee recalled, "We kinda hung on to it."[4] Mendelson received no response at all. Both Clinton political appointees and career employees at the EPA knew that it was wisest to wait for Gore to assume the presidency.

The shockwaves of the presidential election in 2000 turned the EPA upside down. Gore was of course *not* elected president, but it had taken an extra month to sort out the election returns, which left only a few weeks remaining in the Clinton administration before the new president arrived. Carol Browner's EPA responded with a flurry of regulatory activity in Browner's final weeks, days, and hours. Uppermost was publishing in the Federal Register formal findings that would trigger a requirement that the EPA restrict, for the first time, mercury emissions from existing power plants, which, by simultaneously reducing particulate matter emissions, would prevent each year thousands of premature deaths, heart and asthma attacks, and hospitalizations for respiratory diseases, as well as hundreds of thousands of workdays lost to respiratory illnesses.[5]

In a last parting shot, the EPA took action on Mendelson's petition. On January 12, 2001, after sitting on it for more than fourteen months, the EPA's deputy administrator, Robert Perciasepe, signed a formal notice requesting public comment on the petition. Because of normal processing delays between agency action and Federal Register publication, the notice was not published in the Register until January 23, 2001, several days after George W. Bush's inauguration. One outgoing Clinton administration EPA political appointee characterized the Federal Register publication as being akin to "leaving a turd on the door of the next [administration]."[6]

Bush's Pledge

At the very outset of the Bush presidency, environmentalists had reason to be cautiously optimistic about the incoming administration. Or at least reason to hope that any fears they might have harbored of a return to the "Reagan years"—when a newly elected Republican president had pledged to cut back sharply on EPA regulations and had searched for an EPA administrator "willing to bring EPA to its knees"[7]—were misplaced.

After all, Bush had made a bold campaign pledge to address climate change by regulating greenhouse gas emissions, something no president or presidential candidate had previously done. In October, a month before the election, he had vowed to enact sweeping new air pollution controls, including "mandatory reduction targets" for carbon dioxide from the nation's power plants. Lest this effort to outflank his opponent on the left be unsuccessful, Bush expressly contrasted his aggressive regulatory approach with Gore's much weaker promise to advocate for "only a voluntary program."[8]

Following his election, Bush appeared well on track to honor his campaign promise. He named to the most senior positions in his cabinet a series of highly regarded luminaries who openly and enthusiastically supported climate change regulation. They included Secretary of the Treasury Paul O'Neill, Secretary of State Colin Powell, and National Security Advisor Condoleezza Rice. Not all cabinet positions are created equal, and Treasury and State are two of the most prestigious.

Secretary O'Neill was the most outspoken of the three. In 1998 he delivered a speech in which he equated the threat of climate change to the threat of nuclear holocaust and described global warming as a "real danger to civilization."[9] In his interview with President-elect Bush for the Treasury position, O'Neill informed Bush that he was "deeply involved in global climate change."[10] And after his Senate confirmation, Secretary O'Neill proudly put a copy of his 1998 speech "at every place around the cabinet table" at the first meeting of the cabinet with the president.[11] Secretary Powell arranged for a formal briefing on climate science only nine days after taking office, when he was informed of the scientific consensus that "the issue is a real one," "the first signs of human-caused climate change have likely occurred," and further significant and serious changes were "inevitable."[12]

Bush's pick to head the EPA, New Jersey governor Christine Todd Whitman, held equally strong views on the climate issue. Whitman believed that the federal government needed to take aggressive action, including regulating greenhouse gas emissions from power plants, to address climate change.[13] The climate issue was largely the reason she had accepted the president's nomination to serve as EPA administrator, buoyed by Bush's campaign pledge to regulate carbon dioxide from power plants.[14]

At the time when Whitman took the helm at the EPA, on January 31, 2001, the position of EPA administrator may have lacked the gravitas of a "real cabinet" position such as Treasury and State, but Governor Whitman herself possessed enormous national stature. A rising star of the Republican Party, she was widely considered to be presidential material.[15] She had been considered by Bush[16] (and in 1996 by Robert Dole)[17] as a vice-presidential running mate, only to lose out to Dick Cheney.

As New Jersey governor, Whitman had a reputation as being business friendly but also serious about environmental protection, including addressing climate change.[18] She had highlighted the climate issue as early as 1994, and New Jersey announced its Sustainability Greenhouse Gas Action Plan in 1999.[19] A year later it adopted a plan to reduce greenhouse gas emissions originating in New Jersey.[20] In announcing Whitman's nomination to head the EPA, President-elect Bush praised her environmental record and stressed, "She and I share the same point of view."[21] Once in office, Whitman left no doubt that she intended climate change to be her signature issue. And she found ready allies in O'Neill, Powell, and Condoleezza Rice.

As EPA administrator, Whitman wasted no time in putting the climate issue front and center. On February 26, less than a month after taking the helm at the EPA, she appeared on CNN to reiterate Bush's campaign pledge: "Bush was very clear during the course of the campaign that he believed in a multi-pollutant strategy, and that includes CO_2 [carbon dioxide]," she said.[22] After consulting at length with the White House chief of staff, Andy Card, Whitman received what she understood to be the green light necessary for her to stake out a similarly aggressive regulatory position internationally.[23] On March 3, in Trieste, Italy, at the G8 Environmental Ministerial Meeting Working Session on Climate Change, Whitman once again referred to the president's campaign pledge to target power plant carbon dioxide emissions for mandatory reduction targets, emphasizing that power plants were "the largest source of such emissions in the United States." Whitman further made it clear that "the Bush Administration considers global climate change to be one [of] the greatest environmental challenges we face."[24]

What Whitman did not know was that while she was overseas, the political landscape in the nation's capital on the climate issue was dramatically shifting. The green light she had taken great care to receive from the White House before her trip proved meaningless.

"Kneecapped"

On March 13, only one week after Whitman returned from Trieste, the president publicly repudiated his campaign pledge, announced that the EPA would not regulate carbon dioxide emissions, and added for good measure that it lacked the legal authority to do so.[25] Though these policies were central to the EPA's work, Whitman wasn't consulted even once.[26] She was only six weeks into her tenure as EPA administrator, but, as Secretary O'Neill later remarked, "Her career was basically over."[27]

Whitman went through the motions of staying on the job for another two years, but mostly in a ghostlike fashion. The president of the United States had cut her loose unceremoniously, and all of her clout and stature were publicly shredded. She made her mark on a few matters in her remaining time at the EPA, including ordering General Electric to pay to remove PCBs from the Hudson River and a proposal to significantly reduce off-road vehicle diesel emissions, but she never was able to regain any traction on the climate issue or on any other major initiative. After Whitman formally resigned as EPA administrator in July 2003, she never again returned to the national scene, disappearing into political anonymity following the 2005 publication of her book *It's My Party, Too: The Battle for the Heart of the GOP and the Future of America,* which sharply criticized the Republican Party for its increasing conservatism and the Bush-Cheney administration for being unduly divisive.

What happened? The answer is simple. As many others would later repeatedly say, Vice President Dick Cheney "kneecapped" Whitman. It was "a clean kill," wrote Cheney's biographer. "She ended up looking like a fool."[28] Whitman was Cheney's first casualty, but not his last. He was the consummate inside player during his eight years in the White House: outmaneuvering others, including the president, to ensure that his own policy preferences prevailed. Cheney earned the moniker "The Angler."[29]

Immediately after the inauguration, at Cheney's own instigation, President Bush charged the vice president with running the National Energy Policy Development Group, popularly described as the "Cheney Energy Task Force," to "develop a national energy policy designed to help the private sector, and, as necessary and appropriate, state and local governments, promote dependable, affordable, and environmentally sound production and distribution of energy for the future."[30] The clear, yet largely unstated, premise of the Cheney Energy Task Force was that the EPA under Clinton had hampered the nation's energy industry with unduly aggressive application of environmental protection laws, especially in its application of the Clean Air Act to power plants.

For that reason, Cheney did not take lightly Whitman's February 26 comments on CNN, which seemed to double down on the president's campaign pledge to regulate carbon dioxide emissions from power plants. To the vice president, her comments were a clear shot across the bow and a challenge to the mission of his task force. No more pleased were those in energy industry who had supported the Bush-Cheney ticket.

Three days after Whitman appeared on CNN, former Republican governor of Mississippi and utility lobbyist Haley Barbour gave Cheney political ammunition to lobby their shared cause to the president. Barbour wrote in clear and unambiguous terms: "A moment of truth is arriving. . . . The question is whether environmental policy still prevails over energy policy with Bush-Cheney, as it did with Clinton-Gore." Barbour characterized the "Clinton-Gore" approach as "eco-extremism" that had resulted in "less energy and more expensive energy."[31] He focused specifically on "whether this Administration's policy will be to regulate and/or tax CO_2 as a pollutant," leaving no doubt what he thought the answer must be. Unequivocally: no.[32]

It would normally take months for an administration to develop a formal policy on climate change, including the role of regulation of carbon dioxide. And it was far from certain that such a protracted planning process, involving all the interested members of the cabinet, would come to what Cheney and Barbour believed to be the correct answer. Not only would it require an explicit reversal of a campaign pledge, but EPA administrator Whitman was not the only member of the president's cabinet who strongly supported a regulatory approach to climate.

Cheney's responsive tactics were masterful. He orchestrated the transmittal to the president of a formal letter from four prominent Republican senators who were unified in their hostility to climate regulation—North Carolina's Jesse Helms, Nebraska's Chuck Hagel, Kansas's Pat Roberts, and Wyoming's Mike Enzi. Collectively, they requested "a clear understanding of your Administration's position on climate change, in particular . . . the regulation of carbon dioxide under the Clean Air Act." The senators' letter left no doubt about their shared hostility to Whitman's statement on CNN. Indeed, it made explicit reference to those comments.[33]

The letter landed at the White House just as Whitman was landing back in the United States from Trieste. For both O'Neill and Whitman, everything about the letter, its timing and wording, was Cheney. He was the puppeteer, pulling all the strings.[34]

Whitman immediately called the White House chief of staff to request a meeting with the president. The first time she could get was the morning of March 13 at 10 a.m. But by the time Whitman arrived and walked into the Oval Office, it was already a done deal. Before she could begin to make her pitch, the president cut her off: "Christie, I've already made my decision," he said, and he informed her that he had already signed a letter formally responding to the senators' question. The ink on the signature was barely dry. Upon exiting the Oval Office, Whitman watched the vice president put on his coat and pick the letter up from a secretary's desk—after asking her, "Well, is the letter ready?"—he was on his way to deliver it to Capitol Hill.[35]

Condoleezza Rice, having caught wind earlier that same day of what was in the offing, contacted Colin Powell, and they had agreed to go immediately over to the White House to meet with the president for the purpose of blocking the letter. But Cheney's black limousine motorcade, flanked by Secret Service and escorted by a fleet of motorcycled police officers, had already left the White House before Powell arrived at the security gate. Rice later told Powell: "They all got together . . . and said to hell with everybody else and just signed it."[36]

Whitman called O'Neill immediately afterward, furious. The Treasury secretary shared her frustration. "We just gave away the environment," he told her, "for no good reason."[37] She later described it as "equivalent to flipping the bird, frankly, to the rest of the world on an issue about which they felt so deeply."[38]

Neither Whitman, O'Neill, Rice, nor Powell, nor any of their respective staffs, were consulted or saw the president's response letter to the four Republican senators or any draft letter before it was sent. Whitman was convinced that Cheney was the letter's primary architect from start to finish.[39] But regardless of its origins, the letter made official President Bush's complete repudiation of candidate Bush's campaign pledge to regulate carbon dioxide emissions from power plants.

Cheney's carefully crafted letter did far more than answer the policy question posed by the Republican senators. It took a further extraordinary step, one not requested by the senators who had written to the president. Bush's letter announced a legal conclusion: that even if the president of the United States wanted, as a matter of policy, to undertake such regulation, the EPA lacked authority to regulate carbon dioxide under the Clean Air Act. Without consulting with either the EPA or the Department of Justice, the president's letter made an authoritative interpretation of federal law: carbon dioxide "is not a 'pollutant' under the Clean Air Act."[40]

By taking that additional step, the president had crossed a formal line. He had taken the position, not merely that the EPA should not regulate greenhouse gases as a matter of *policy*, but that the agency lacked the authority to do so as a matter of *law*. The president has the power as the chief executive to interpret federal statutes enacted by Congress, and this was what President Bush had now done in his letter to the senators. Presidents have wide-ranging policy discretion in deciding whether, when, and how to exercise the authority conferred on them by the Constitution and by Acts of Congress. But as the U.S. Supreme Court made clear in no less than its landmark 1803 ruling in *Marbury v. Madison* almost two centuries before, "It is emphatically the province and duty of the judicial department to say what the law is."[41] Bush had, in effect, the *first* word about whether the EPA possessed legal authority to regulate carbon dioxide and other greenhouse gas emissions. But the Supreme Court would have the *last* word.

The battle lines leading to *Massachusetts v. EPA* had therefore been drawn. As Haley Barbour had forecast, the "moment of truth" had indeed "arrived."[42]

4

An Agency Misstep

President Bush's letter had, in effect, dictated the EPA's response to Mendelson's petition. Because, by presidential decree, carbon dioxide and other greenhouse gases were not "pollutants" within the meaning of the Clean Air Act, the EPA's mandate was now clear: the agency could not regulate greenhouse gas emissions. Although the president had not consulted with the EPA or any of its lawyers before announcing his legal conclusion, neither the EPA administrator nor any other EPA official, including its general counsel, had authority to contradict the president. They either had to accept Bush's judgment or resign. As Marianne Horinko, who became acting EPA administrator upon Whitman's resignation in July 2003, put it, "If Whitman couldn't win the battle, then I sure as hell . . . [was] not going to run up there and take a crack at Dick Cheney."[1]

The question remained, however, what, if anything, to do with the petition. Should Whitman just sit on it, as the Clinton EPA had done until its final days? Or should the EPA formally deny the petition now that the president had staked out his position? Because no federal law had expressly provided for the filing of Mendelson's petition in the first instance, or even suggested the possibility of such a filing, no federal law established a specific time frame within which the EPA had to respond to such a petition. The Clinton administration had ignored Mendelson's essentially specious claim that the EPA had to respond to his petition within 180 days. The Bush administration could now do the same.

Ironically, it was the Bush political appointees most opposed to climate change regulation who decided they could use Mendelson's petition to their political advantage, to cement the president's position as formal EPA policy. Absent that political calculation, the Mendelson petition might well have effectively disappeared. But their calculation ultimately proved to be a *mis*calculation that would be costly to their desired policy ends years later at the Supreme Court.

"Do-Nothing"

In the aftermath of Bush's letter to the senators, the lawyers in the EPA's Office of General Counsel knew they had their marching orders. In any official statement on the extent of the EPA's authority to regulate greenhouse gases, the agency attorneys would have to say, consistent with the president's declaration, that the EPA lacked authority because greenhouse gases were not "air pollutants" within the meaning of the Clean Air Act. That was a given. But they also knew that any such legal conclusion would be subject to an aggressive attack in litigation and would therefore need to be backed up by persuasive legal argument. Unlike executive branch lawyers, federal judges would not just defer to a president's say-so.

The EPA attorneys were well aware that in crafting a legal argument, they faced a threshold hurdle. They would need to repudiate the legal opinion of the two prior EPA general counsels who, during the Clinton administration, had in a formal general counsel opinion and in congressional testimony taken the precise opposite view: each of those general counsels had concluded that carbon dioxide and other greenhouse gases *were* Clean Air Act air pollutants.[2]

Such agency reversals are not as rare as one might suppose, and they are not inherently improper. The Supreme Court, in its leading case on when courts should defer to a federal agency's reading of a statute it administers, *Chevron v. Natural Resources Defense Council, Inc.* (1984), upheld an agency reading of a statute that differed from the reading given that same statute by the same agency years earlier during a different presidential administration. The Court in *Chevron* explained that there is nothing necessarily untoward in the agency's reaching that new interpretation and rejecting the old one. There can be legitimate reasons, the Court reasoned, for a new administration to strike a different balance between competing policy preferences in its reading of the federal statutes it administers. Politics and elections can, the Court underscored, matter.[3]

Still, such an agency reversal raises an obvious red flag. Just because an agency can validly change its mind does not mean it acts validly every time it does so. That such an agency reversal is not automatically *invalid* does not mean it is automatically *valid*. At the very least, the agency must offer a persuasive explanation for why it was wrong before and not wrong this time. Or, if not wrong before, it must explain what had changed between the first interpretation and the second that now makes the latest lawful. For the Bush EPA, this would mean offering a persuasive explanation for why the EPA had changed its mind about how best to read statutory language that itself had not changed since two EPA general counsels had concluded, based on extensive legal analysis, that carbon dioxide was an "air pollutant." Potentially doable? Yes. But necessarily persuasive? No. Good lawyering on behalf of the president's action would be needed.

The best lawyering is generally done *before* a major decision is made, so that potential legal problems can be addressed and thereby minimized before they arise. And this is where Cheney unwittingly outflanked himself in his haste to prevent Whitman, Rice, Powell, and O'Neill from persuading the president not to send the letter. By circumventing Whitman and rushing a presidential decision, Cheney failed to allow for any consultation with the expert lawyers at the EPA. Had he done so, they would have explained to Cheney (and the White House) that the legal question the president's letter purported to answer—whether carbon dioxide was an air pollutant—was not, as the letter assumed, binary in nature, meaning susceptible to being answered simply yes or no. The actual "legal" answer they needed, in order to effectuate their preferred policies, was more nuanced—sometimes yes and sometimes no, and not the unqualified no expressed in Bush's letter to the senators.

True, the Bush administration did not want to consider greenhouse gas emissions as an "air pollutant" for *regulatory purposes*—providing the EPA with authority to restrict them—but the president had made it clear in other proposals he had made that he did want to continue to treat carbon dioxide as an "air pollutant" for *research purposes,* thereby allowing the EPA to continue to spend sums authorized by Congress for studying "air pollutants," both to consider the impacts of greenhouse gases on public health and welfare and to develop voluntary initiatives to reduce greenhouse gas emissions.[4] The president's letter did not, however, appreciate this legal nuance.

As a result, the letter's simple answer that carbon dioxide was not an air pollutant at all risked undermining the agency's authority to engage in the climate change research that the administration not only supported but also was arguing was necessary before the EPA could reasonably make a decision as to whether to mandate restrictions on greenhouse gas emissions. If the EPA lacked any authority to study climate change—because greenhouse gases were not "air pollutants"—the administration's reasoning no longer made any sense.

Whitman accordingly faced a practical problem: how to square what the president's letter stated with the administration's actual climate policies. The agency's initial response was the tried-and-true tactic of most organizations confronting a difficult problem with no clear deadline: do nothing. So the EPA continued to sit on Mendelson's petition and did not formally act on it at all.

There were good strategic reasons why many career EPA employees, including attorneys in the Office of General Counsel, quietly favored this "do-nothing" approach. Not surprisingly, most career EPA employees tend to believe in their agency's mission to protect the environment. After all, that is why they decided to become environmental lawyers, environmental scientists, environmental engineers, or environmental economists working in public service. They believe in the importance of environmental protection and in the ability of the EPA and other aligned agencies to do good work. It would be the odd person who would decide to become a career public servant at the EPA without harboring a strong desire to enlist the agency's authority to address the nation's environmental ills.

Beyond that, the notion that the agency possesses legal authority is far more attractive than the proposition that it does not. Few who dedicate their professional lives to address a problem about which they care deeply are eager to conclude that they lack the authority to tackle the problem in a meaningful way. This was especially true for the EPA employees who were trying to decide what to do about Mendelson's petition in the immediate aftermath of the president's letter to the senators. Because career EPA employees are just that—*career* employees—many had worked on the prior EPA general counsel opinions that the president's letter had repudiated.

That is likely why career EPA employees' preferred response was to decline to address Mendelson's petition at all. Just let it sit. To respond would require the EPA to take a formal position as an agency. So far the president had taken a position, but the EPA had not; the only time it had taken a formal position, it had embraced the opposite view in favor of its legal authority. Should the EPA act now, the agency would necessarily have to embrace the president's legal conclusion and formally reject its prior legal interpretations. All else being equal, career EPA employees came to the conclusion that it would be best to stay quiet, awaiting the next presidential election cycle and the arrival of a new president who might favor federal climate change regulation.[5]

"A Big Favor"

Two other factions favored pressing the issue, for diametrically opposed reasons. The first consisted of hard-line political appointees within the EPA who, unlike Whitman and her general counsel, shared Vice President Cheney's view that the EPA had overreached its authority during the Clinton administration. The second consisted of environmentalists outside of the EPA who, like Mendelson, had tired of the EPA's persistent failure (under both Democratic and Republican administrations) to take action to address the threat of climate change in a significant way by restricting greenhouse gas emissions. They were no less tired of waiting for the "right time" and wanted the courts to step in. But to have that happen, they needed the EPA to act on the petition.

Leading the charge for the first faction was Jeffrey Holmstead, a presidential appointee who was serving as assistant administrator for air and radiation. Forty-one years old at the time, with sandy-brown hair and the athletic build of a wrestler, Holmstead came to the EPA with a distinctly impressive record of academic, personal, and professional achievement. Born in Utah and raised in Boulder, Colorado, Holmstead served as a missionary for the Mormon Church for two years in Argentina before graduating first in his class from Brigham Young University in 1984, where he sang bass in the university choir. He then attended Yale Law School, where he joined the newly formed conservative Federalist Society and continued to excel academically. Upon graduating in 1987, he secured a judicial clerkship working for Judge Douglas Ginsburg on the United States Court of Appeals for the District of Columbia Circuit: one of the most highly competitive clerkships in the nation.[6]

After a short stint at a D.C. law firm, Holmstead joined George H. W. Bush's White House Counsel's Office in September 1989, first as assistant counsel to the president before being promoted a year later to associate counsel.[7] At the White House he worked primarily on environmental issues, including the administration's effort in 1990 to secure passage of the Clean Air Act Amendments. Holmstead left the White House in January 1993 and joined the D.C. office of Latham Watkins, a highly regarded national law firm, where he represented industry, including electric utilities, on Clean Air Act issues.[8] When Holmstead had entered law school a decade earlier, he had planned on a corporate law career and did "not even know there was anything called environmental law." But by the close of the first Bush administration, he had decided to devote his career to environmental law.[9]

During his 2001 confirmation hearing to be assistant administrator of air and radiation at EPA, Holmstead properly stressed that he was "very eager" to work closely with Administrator Whitman.[10] But insiders knew that Holmstead owed his new position at the EPA not to Whitman but to persons in the White House and private industry who sought to cut back on what they saw as overly aggressive Clean Air Act enforcement by the EPA during the Clinton administration. When Whitman's preferred appointee to be her deputy administrator, Linda Fisher, tried to shunt Holmstead off to the side to work on water rather than air issues, the White House agreed to let Fisher have the deputy job only if she would agree to Holmstead's serving as the head of the EPA's Air Office. At the time when Cheney was working to undercut Whitman on the climate issue in March 2001, Holmstead (unlike Whitman) knew from his own sources at the White House that the president's letter to the senators would disclaim any Clean Air Act authority over greenhouse gases.[11]

Holmstead's interests differed markedly from those of the EPA career employees. He wanted to repudiate the prior EPA general counsel interpretations and cement a narrower view of EPA authority. Just as career employees might prefer to keep the old general counsel opinions on the books, Holmstead sought to be rid of them. And he sought to do so on sweeping grounds that would make it difficult for future EPA administrators to resurrect them. He wanted, as he put it, to "settle the question once and for all."[12] And a means of achieving this was readily at hand: a formal denial of Mendelson's petition.

Mendelson, too, wanted a formal decision from the EPA, but to achieve very different ends. He now knew it was preordained that the agency would deny his petition. And that was fine, because his goal was to litigate. The EPA was clearly not going to do anything. So environmentalists' best shot, to his mind, was the courts. And it would be far easier to challenge a formal agency decision than the absence of a decision. Toward that end, on January 10, 2002, Mendelson formally notified the EPA by letter that unless the agency responded to his petition by June 1, he would initiate litigation to compel the agency to act.[13]

Mendelson's threat was not a complete bluff, but it lacked much oomph. In theory, a plaintiff can bring a lawsuit under the federal Administrative Procedure Act to compel a federal agency to take action that has been "unlawfully withheld or unreasonably delayed."[14] But federal courts are understandably very reluctant to order an agency to do so in the absence of a specific statutory directive that a federal agency must take certain kinds of action and within certain prescribed time limits. Courts appreciate that federal agencies invariably have far more on their plate than they can possibly handle, and generally believe that agencies are better positioned than a court to determine how best to allocate an agency's limited resources to the most pressing priorities. The mere fact that one person bringing a lawsuit might prefer that the agency address some issues before others is not much of a reason for a court to second-guess the agency's decision to allocate its scarce resources differently. That is why lawsuits brought against agencies for unreasonable delay almost never secure any effective relief. At best the court might compel the agency to respond by explicitly saying "we're really busy doing other important things."

Almost a year after Mendelson's notice letter, the EPA had still not responded. Mendelson had to decide whether to take the next big step and file a formal complaint in court to compel the EPA to act based on "unreasonable delay." While waiting for the EPA to act, he had mostly been working for the past couple of years on food safety matters related to genetically modified organisms. He was personally and professionally ready to return to pushing the EPA on the climate issue, but he was worried that he lacked the financial resources to maintain the lawsuit. The Energy Foundation had previously given him a small grant to support his earlier efforts, but that was only enough to set things in motion, not to litigate the case itself. Litigation would be much more expensive, and given the opposition of other environmentalists to his past efforts, he could not be at all confident that he could secure the necessary funds. Just the opposite. There was reason to be concerned that a lawsuit could further erode his organization's funding.[15]

When some environmentalists in the powerful national organizations had previously caught wind of Mendelson's intention to push the EPA to make a decision on his petition, they lobbied him hard to stand down. Even the charitable foundation that supported his work pulled its grant, most likely because other environmental groups, using hardball tactics, contacted them to complain. The organization was on a thin budget at best, and the loss of the grant was a devastating blow. Highly placed environmental attorneys in leading organizations called Mendelson and pressed him: "Don't do this."[16]

The gist of the dispute was where and how best to fight the Bush administration on the climate issue. The powerful national groups like the Natural Resources Defense Council preferred to litigate the issue in California in defending California's efforts to regulate greenhouse gas emissions in the federal court of appeals out west. They were already playing a central role in that litigation, were better positioned to control the arguments being made, and considered the federal courts in California a more likely place for a favorable judicial ruling. The Sierra Club's leading climate attorney, David Bookbinder, had witnessed the aggressive opposition to Mendelson's pressing his own case. "Oh my God, sweet Jesus, they went crazy, they went batshit over this idea." The resulting pressure on Mendelson, Bookbinder said, was "incredible."[17]

Bookbinder privately broke from the pack and behind the scenes provided Mendelson with the assist he desperately needed. Bookbinder had met Mendelson a few years earlier and had been impressed by both him and his petition. When Bookbinder's own boss, Sierra Club president Carl Pope, rebuffed Bookbinder's suggestion that the Sierra Club support Mendelson's filing of a lawsuit—"No, we're not gonna do this"—Bookbinder crafted an effective end run. He first persuaded Pope to agree that the Sierra Club would support Mendelson if "they're gonna do it on their own, in which case we have to join them," and then immediately called Mendelson and instructed Mendelson to tell him "you're gonna do this alone anyway," because if you do "I've gotta do it with you." Mendelson quickly obliged—telling Bookbinder "we're gonna do this alone"— and Bookbinder happily responded "Fine, let's file this." Let's, he said, "get the goddam answer" from the EPA.[18]

One last-ditch effort was mounted to stop the lawsuit. At 11 p.m. on the night before Mendelson was planning to file, the highest-ranking career EPA employee who had been working closely with Administrator Whitman in favor of climate regulation, and who no doubt had been clued in to Mendelson's plans by the Natural Resources Defense Council, privately called him at home. The caller's off-the-record message was the one he had heard before: "Don't do this," he said. This "is not the way to do business in D.C." The lawsuit, he was told, will undermine the ability of the folks at the EPA who believe in addressing the climate issue to take action.[19]

The very next day, Mendelson filed his lawsuit anyway.[20] His daughters were no longer a baby and a toddler. Anna was in first grade, and Quincy would soon be in prekindergarten. The climate clock was ticking, and the time for further waiting was over. Mendelson had the support of his own organization and his family to follow through on what he had set in motion three years earlier. He had no interest in acquiescing to the way "business in D.C." had long been done. What he wanted was for business to be done differently. Mendelson delivered his complaint to the U.S. District Court for the District of Columbia on December 5, 2002, requesting that the court compel the EPA to act on his petition. And this time he was joined by the Sierra Club. Apart from Greenpeace, a proudly more provocative group that has always been willing to go on its own, no other major environmental organizations joined the lawsuit.[21]

The EPA Office of General Counsel contacted Mendelson soon thereafter. They wanted to talk settlement. A meeting of the minds was relatively easy to reach because both Mendelson and the EPA political appointees like Holmstead wanted the same thing: a formal agency answer to the petition. The EPA made it clear that it would respond to the petition within the next several months, at which time the parties would agree to dismiss Mendelson's lawsuit and Mendelson would be free to challenge the agency's formal decision in court.[22]

And that is exactly what happened. After months of internal wrangling within the EPA between the agency's career attorneys and the new administration's political appointees, on August 28, 2003, the EPA publicly released two documents. The first was a formal opinion by Robert Fabricant, the EPA's general counsel, which provided that he "formally withdraws" the opinion of previous EPA general counsel Jonathan Cannon regarding the status of carbon dioxide and other greenhouse gases under the Clean Air Act.[23] The second was a formal agency decision denying Mendelson's petition.[24]

Following lockstep behind the president's letter and its architects, the EPA political appointees joined Cheney in stepping in the turd left on their doorstep by the Clinton administration. As summed up years later by one of the environmental attorneys who later joined Mendelson in challenging the EPA's action in court, the EPA "did us a big favor by actually denying the petition."[25]

"Jeff Has Done the Deed"

A careful reader of the new EPA general counsel opinion would have noticed that it stopped short of embracing the president's March 2001 letter to the Republican senators in its entirety. The letter had baldly and unqualifiedly stated that carbon dioxide was not an air pollutant.[26] Fabricant's opinion was more nuanced. Unlike the White House, which had prepared the president's letter without first consulting legal experts, some lawyers at the EPA had clearly been at work in helping Fabricant draft his opinion. Holmstead later acknowledged he "was pretty involved in putting together" what he described as "the so-called Fabricant memo" because Fabricant was "more of a counselor to Whitman" and, unlike Holmstead, "was not a Clean Air Act lawyer."[27]

The Fabricant opinion (regardless of its author) stated that the general counsel had "determined that the [Clean Air Act] does not authorize the EPA *to regulate for global climate purposes*" and that "accordingly, CO_2 and other [greenhouse gases] cannot be considered 'air pollutants' *subject to the [Clean Air Act's] regulatory provisions.*"[28] In other words, Fabricant was quietly acknowledging that carbon dioxide and other greenhouse gases could be considered "air pollutants" for purposes other than imposing mandatory limits on greenhouse gas emissions to address climate change. Under Fabricant's opinion, the EPA could therefore treat greenhouse gases as "air pollutants" based on their contribution to climate change when deciding to fund climate research and for the purpose of supporting federal programs aimed at promoting industry's *voluntary* reduction of greenhouse gas emissions. But it could not treat greenhouse gases as air pollutants for the purpose of establishing federal regulations directly restricting their emissions by sources such as motor vehicles or power plants to reduce climate change.

Fabricant's opinion allowed the EPA to regulate the emissions of air pollutants that were greenhouse gases under the Clean Air Act so long as that regulation was not "for global climate change purposes."[29] This was an important and necessary concession because the EPA had long been doing just that. For instance, methane is a powerful greenhouse gas, but it is also an air pollutant that can cause serious respiratory illnesses. The EPA had for years acknowledged its authority to regulate methane emissions under the Clean Air Act, but only based on the adverse health impact of an individual's breathing in methane and not for "climate purposes."[30] The Fabricant opinion was, without formal acknowledgment, taking care to preserve the EPA's authority to regulate methane and other gases in these other non-climate contexts.

One might think that these nuances in the Fabricant opinion were of little import. After all, the bottom line of the president's letter and the Fabricant opinion was the same: no regulation of greenhouse gases, including carbon dioxide, under the Clean Air Act. But they were in fact quite significant. Fabricant's necessary concessions had exposed significant weaknesses in the legal arguments in support of the president's opinion. It made it clear that the validity of the president's opinion turned on the unlikely proposition that the same chemical compound could be considered an air pollutant for some Clean Air Act purposes but not for others.[31] Carbon dioxide was not an air pollutant for regulatory purposes, but it was for research purposes. Similarly, methane could be regulated for some purposes, but not for others.

As a matter of abstract policy, both propositions were plausible: the EPA's (and by extension the president's) problem was that there was no obvious support in the relevant statutory language. Each proposition was hard to square with the fact that the Clean Air Act provided a single definition of "air pollutant." And that definition expressly included "any chemical compound" that was "emitted into . . . the ambient air."[32] That would encompass carbon dioxide, methane, and all other greenhouse gases. The EPA's lawyers, in other words, unlike the White House, knew they had a problem.

The second EPA document, produced on the same day as the Fabricant opinion, was the agency's formal denial of Mendelson's petition. Here too, the bottom line was unequivocal, mandated by the president's March 2001 letter. But the structure and precise wording of the agency's response suggested an internal EPA awareness of potential weaknesses in the president's legal position.

First, oddly, the administrator of the EPA did not herself sign the petition denial. Administrator Whitman did not sign it for an obvious reason: she was no longer EPA administrator. She had resigned from the administrator position only weeks before.[33] Given her opposition to the president's letter and its underlying legal position, there is good reason to suppose that the timing of the petition denial was not mere happenstance. But even the acting EPA administrator, Marianne Horinko, did not sign the denial, although it would have been routine practice for the acting administrator to do so.[34]

Instead, the denial was signed by Holmstead, in his official capacity as assistant administrator for air and radiation. Although it is unusual for the denial of a petition directed to the administrator to be signed by a lower-ranking political appointee, it was fitting that Holmstead put his personal stamp on the decision. More than any other political appointee, he had spearheaded the position that the EPA lacked the authority to regulate greenhouse gases under the Clean Air Act.[35] For that same reason, in another departure from the norm, he had had a major hand in drafting the opinion. Custom would have been for career attorneys and political appointees in the EPA Office of General Counsel to do the drafting. In this case, the task had been led by Holmstead.

The EPA did not rely only on the general counsel opinion as the legal basis for denying Mendelson's petition. The agency hedged its bets with a backup argument, as any good lawyer would have recommended. Sure, if your first argument is solid, go for it. But if it might turn out to be a loser, then be ready with Plan B just in case. It's the lawyer's equivalent of spreading your eggs among more than one basket.

The EPA denial left little doubt that those who drafted it were worried that their primary legal argument—in support of the president's (and Cheney's) letter—could well be a loser in court. The drafters spent almost as many words (about 4,300) describing Plan B as it did defending the president's position that the EPA lacks authority to regulate greenhouse gases to address climate change (about 4,500 words). So, what was Plan B? Simply put, it was "You can't make me do it." The agency's backup argument was that even if the EPA did possess authority to regulate greenhouse gases to address climate change, the agency had "no mandatory duty"[36] and did not believe it was "appropriate at this time"[37] to regulate.

Although presented as a backup to the primary argument, this Plan B was no mere make-weight. It was clearly the stronger of the two arguments. The absence of anything in the Clean Air Act that imposed a timetable on the EPA's consideration of whether greenhouse gases from new motor vehicles endanger public health and welfare was the Achilles' heel of Mendelson's petition when he first drafted it in 1998. Mendelson knew it then, and the EPA's lawyers knew it too.[38] Arguments that might not sound good in politics because they seem to be based on "legal technicalities" rather than principled policy—and "You can't make me do it" would easily fall into that category—may nonetheless be winning legal arguments.

The EPA's backup argument was pretty straightforward. The agency pointed out that the Clean Air Act mandated regulation of a type of air pollutant emitted by new motor vehicles only after the EPA administrator had made the "judgment" that those emissions endangered public health and welfare, and the administrator had not yet made that judgment. Nor, as the EPA further argued, does the Clean Air Act "impose a mandatory duty on the Administrator to exercise her judgment" at any particular time. The act provides no "specified deadline" for the "discretionary exercise of the Administrator's judgment." Accordingly, even if greenhouse gases were Clean Air Act pollutants, it was still up to the administrator to decide when it was appropriate to make the necessary inquiry into the risks presented by motor vehicle emissions of greenhouse gases and whether those risks met the statutory threshold of "endangerment to public health" necessary to trigger regulation. In this instance, the EPA was deciding nothing more than that the time had not yet come to undertake those inquiries.[39]

In denying Mendelson's petition, the EPA offered a laundry list of reasons it was not yet the appropriate time. The agency questioned whether it "makes sense to regulate" emissions from motor vehicles. The agency described the existing "uncertainties in our understanding of the factors that may affect future climate change." And the agency suggested that greenhouse gas emission restrictions applicable to new motor vehicles might interfere with "fuel economy standards" for motor vehicles already promulgated by the Department of Transportation to promote energy conservation. The potential for such interference derived from the fact that the best way to control greenhouse gas emissions from motor vehicles is also fuel economy standards, which means that if the EPA were to regulate motor vehicle greenhouse gas emissions, there would be two competing federal agencies—the EPA and the Department of Transportation—purporting to regulate the same thing. Finally, the EPA expressed concern that if it were to regulate greenhouse gas emissions from new motor vehicles before other countries had agreed to do the same, the decision could undermine the United States' leverage in then-ongoing climate negotiations with those other nations to reduce their own domestic greenhouse gas emissions.[40]

In short, in arguing that it was not time to act, the EPA threw its own kitchen sink back at Mendelson's idea that it was the right (or at least statutorily compelled) time to decide whether new motor vehicle emissions of carbon dioxide and other greenhouse gases endangered public health and welfare. The agency did not single out any specific reason or subset of reasons as sufficient to justify denial of Mendelson's petition on that ground. The petition denial never claimed that the agency lacked the scientific basis for concluding that greenhouse gas emissions from motor vehicles endangered public health and welfare. It instead listed a lot of reasons it was not the right time to decide the endangerment issue. It closed by saying "*In light of the considerations* discussed above, the EPA would decline petitioners' request to regulate greenhouse gas emissions even if it had authority to promulgate such regulations."[41]

A classic lawyer's gambit. Dodge the real question of whether greenhouse gas emissions from motor vehicles endanger public health and welfare. And throw in a lot of reasons you are not answering the question—without making it clear which is the most important or the strongest—and hope one of them sticks. No doubt this strategy made sense at the time for the EPA lawyers who drafted that language. But as they would discover three years later during oral argument before the Supreme Court, not everyone was impressed.

Holmstead signed the petition denial at 1:49 p.m. on August 28, almost immediately after the EPA general counsel, Robert Fabricant, had issued his formal revocation of the former EPA general counsel opinion.[42] Fabricant had initially opposed the issuance of the new opinion, advising his staff that it wasn't his job as general counsel to interpret away the agency's authority, but he was overruled by Holmstead and others.[43] Fabricant had quietly submitted his letter of resignation two weeks earlier, and he left the EPA less than a week after Mendelson's petition was denied, with no clear plans beyond returning to the private practice of law in New Jersey. Like his mentor (former EPA administrator Whitman), he then disappeared from the national scene, joining a series of law firms before becoming counsel to a New Jersey business that manufactured building material from contaminated dredged sediment.[44]

Upon learning that the petition denial was now official, a career attorney waited only seconds before emailing her career colleagues in the Office of General Counsel: "Jeff has done the deed!"[45] She followed up later in the afternoon with a second email to the entire Office of General Counsel as well as the Office of Air and Radiation: "It's signed, sealed and delivered. Next stop—D.C. Circuit."[46]

Everyone knew. The preliminary skirmishes and formalities were over. Thirty-seven months had passed since Joe Mendelson had filed his petition with EPA in October 1999. The litigation was now, finally, ready to begin.

5

The Carbon Dioxide Warriors

The practice of environmental public interest law could be a tough—and lonely—business. When Joe Mendelson had walked from his office to the EPA's headquarters in 1999 to hand-file his petition, he had been alone. No state attorneys general had supported him. Except for Greenpeace, which has always relished its role as provocateur, the nation's most powerful environmental groups had repudiated his effort. Some had even sought to impede his work.[1]

By late August 2003, however, Mendelson was no longer alone. When it became clear that the Bush EPA was on the cusp of denying his petition, his once-quixotic plan had attracted lawyers from all over the country, including some who had previously placed hurdles in his path. The Sierra Club had been the first to break away and joined Mendelson in late 2002, but by the end of summer 2003, what had been a grudging trickle had become a cascade of support. The emails and phone calls flooded in, with environmental organizations and state attorneys general competing for his attention, beginning with Massachusetts, New York, and Connecticut. Everyone wanted a piece of Mendelson's case. He now found himself in frequent meetings and on conference calls with the nation's most prominent environmental public interest lawyers and attorneys from many of the country's most economically and politically powerful state attorneys general's offices.

Any plans to try to work constructively with the Bush administration on climate change policy had been abandoned soon after the president had reneged on his campaign promise to regulate greenhouse gas emissions and sidelined EPA administrator Whitman. All the environmental groups agreed it was time to litigate and that the EPA's denial of Mendelson's petition would be their best opportunity. As was made increasingly clear by the latest reports of climate scientists, the threat of climate change was too great to allow for more delay.

They dubbed themselves "The Carbon Dioxide Warriors" because carbon dioxide is the greenhouse gas emitted in the largest amounts.[2] Their shared goal was to overturn the EPA's denial of Mendelson's petition—and to do so, they knew that they had to work together. Coordination was necessary to provide strength through numbers and to avoid the possibility of individuals advancing inconsistent, conflicting legal arguments in litigation that might, unwittingly, undermine their common cause. But as they would soon learn, coordination is no cakewalk when those seeking to coordinate are strong-willed lawyers with competing professional ambitions, personal egos, and ideas.

Five Guys

The now-large litigation planning team first met with Mendelson on September 3, 2003, after the EPA had announced its decision to deny his petition but before it had published that formal denial in the Federal Register, an official daily publication in which executive branch agencies provide notice to the public of their proposed and final decisions. There were too many people to meet in Mendelson's own small office, so they met instead in the Sierra Club's Washington, D.C., office building, a handsome three-story brick building located a few blocks from the Supreme Court. Some were there in person; more than twenty-five governmental and nongovernmental organization attorneys from across the country joined in by conference call.

The team continued to meet in person and by conference calls seven more times in September through mid-October, with their numbers shifting as various organizations, states, and local governments opted in and out of the litigation. The participants discussed logistics, procedural hurdles, and competing legal arguments for overturning the EPA's denial of Mendelson's petition so they would be ready to launch their coordinated, simultaneous attack—with both press releases and legal filings—on the same day and, as required by law, no less than thirty and no more than sixty days after the Federal Register publication date.[3]

There was no single leader of the group, though five attorneys quickly assumed the most prominent leadership positions. Naturally, this included Mendelson, because he had written and filed the petition that would soon make litigation against the EPA possible. The other four attorneys represented either states or leading national environmental groups. They were not a diverse lot in terms of their backgrounds and demographics. All five were white men. All, except for Mendelson, had attended Ivy League schools, either as undergraduates or for law school.

David Doniger

One of the five stood out. The Natural Resources Defense Council's David Doniger was the dominant personality in all the strategic discussions and litigation decisions. Tall, thin, with arching eyebrows and a graying beard, and in his early fifties, Doniger was widely considered the environmental public interest community's foremost expert on the Clean Air Act. A graduate of Rye Country Day School from a wealthy New York City suburb, and Yale University, Doniger pivoted to law school and environmental law after an aborted interest in urban planning.[4]

He had found urban planning toothless and "indescribably boring," but his master's degree in urban planning studies in the early 1970s had exposed him to the recently enacted federal Clean Air Act, which, he concluded, had the potential to force improvements in urban air quality. The Clean Air Act became Doniger's primary focus once he began his legal studies at Berkeley Law, though the absence of any environmental law courses at the time meant he had to learn the subject largely on his own. He worked on air pollution issues for the Natural Resources Defense Council the summer after his second year of law school and joined the organization full-time in 1978, one year after graduation.[5]

Twenty-five years later, in the fall of 2003, Doniger was perched as NRDC's director of climate policy and no less committed to working on air pollution issues, but his focus had narrowed to a particular class of air pollutants: the greenhouse gases causing climate change. He had remained at NRDC for his entire career, except for an eight-year hiatus at the EPA during the Clinton administration, where he had served as counsel to the head of the EPA's clean air program, as director of climate change policy, and as part of the U.S. delegation to the Kyoto Protocol climate negotiations.

Doniger's depth and breadth of expertise with the Clean Air Act and climate change were widely acknowledged, and matched by his self-confidence. After decades in the nation's capital, including his eight years inside the EPA, Doniger believed strongly in his own judgment as to strategy and in his ability to craft the strongest legal arguments. Nor did he shy away from confronting those who disagreed with him. Or from working behind the scenes to build coalitions to overcome bureaucratic obstacles. Doniger could "make nice" or not, be inclusive or not, be highly respectful or not, depending on what was needed to get the job done. As described by

Doniger, "environmental policy is not for weenies." "You don't win your argument just by stating your arguments. You have to be persuasive and stick to it."[6] Working in close proximity to Doniger could be bruising—as many in government, regulated industry, or allied environmental public interest organizations could attest. None, however, denied his sincerity, effectiveness, or smarts, and his views were justifiably highly respected.

Some speculated that Doniger had cleverly orchestrated the events that had led the Clinton EPA to formally endorse his view that the agency possessed authority under the Clean Air Act to regulate greenhouse gases. While working at the EPA during the Clinton administration, Doniger had written the internal memoranda (prepared without Administrator Carol Browner's knowledge) that, once leaked to the national press, had led to Representative Tom DeLay's confrontation with Browner.[7] Browner's colleagues wondered whether Doniger, frustrated by the lack of EPA action on the climate issue, had deliberately leaked his own memo to the press to provoke the confrontation with DeLay.[8] If true, Doniger's plan smacks of both insubordination and effectiveness. True or not, the speculation itself underscores Doniger's reputation for skillful bureaucratic infighting, a trait he brought to the litigation.

Jim Milkey

Jim Milkey of the Massachusetts Attorney General's Office was also among the core group of five who championed the litigation against the EPA's denial of Mendelson's petition. He had been on the lookout for a high-profile climate change case to bring against the Bush administration when he decided to join Mendelson.

Tall, lanky, sprightly, in his late forties, and beginning to gray like Doniger, Milkey had devoted his entire professional career to environmental law and exuded self-confidence. But unlike Doniger, Milkey had worked exclusively for state government. An honors graduate of both Harvard College and Harvard Law School, with a master's degree in urban planning from MIT, Milkey could easily have obtained a far higher paying job with a law firm following his law school graduation in 1983. Or, like many of his classmates, he could have secured a position with the federal government, which both paid better than state government and was the more traditional "elite" career track for Harvard Law graduates.[9]

Milkey instead chose public service with state government. After law school, he decided to stay in Massachusetts rather than return to his home state of Connecticut, where he had been born. He clerked for a Justice on the Massachusetts Supreme Judicial Court and immediately thereafter he joined the Massachusetts Attorney General's Office, where he had remained ever since. By the time the EPA denied Mendelson's petition in late August 2003, Milkey had worked in that same office for almost two decades and risen to head the attorney general's environmental law division. He was proud of his almost twenty years of work in the Attorney General's Office but was increasingly restless and looking for new opportunities to make a difference.[10]

A high-profile lawsuit challenging the EPA's denial of Mendelson's petition was just such an opportunity, and Milkey's involvement was no mere happenstance. He did not previously know either Mendelson or Doniger. Milkey deliberately inserted himself into the litigation after learning about the case from a Sierra Club attorney.

Several years earlier, Milkey had taken a leave of absence from his work to live in Denmark with his wife, Cathie Jo Martin, a Boston University political science professor. During that year abroad, based on his experiences in Denmark, Milkey decided that when he returned to work at the end of his leave, he would make his professional mark by promoting more demanding restrictions on greenhouse gas emissions.[11]

Soon after Milkey had arrived in Denmark in the fall of 2000, Northern Europe was subject to enormous storms and flooding of historic proportions.[12] It was the wettest autumn in parts of Northern Europe since records were first kept in 1766, with damaging cyclone-level winds approaching one hundred miles per hour and massive flooding. Government leaders and the general public immediately linked the ferocity and persistence of the storms to climate change, which scientists later confirmed.[13] Many quickly blamed the United States, because it had contributed far more greenhouse gases to the atmosphere than any other nation in the world—24 percent of all greenhouse gases, although the United States accounted for only 4 percent of the world's population.[14]

When, a few months later, the newly elected U.S. president, George W. Bush, retreated in March from his earlier promises to regulate greenhouse gases, his actions were swiftly condemned in Denmark and throughout Europe. One opinion editorial in a European newspaper at the time made clear the degree of disdain: "The trademark cowboy boots are the giveaway. George 'Dubya' Bush's decision to put cash before the global environment and ditch the Kyoto Protocol has incensed green campaigners and governments across the planet."[15]

What were still at best back-page stories in the United States were front-page stories in Europe, where Milkey was pondering his future. During Milkey's last month there, when Bush arrived in neighboring Sweden in June, he was met by twenty-five thousand protestors, many of whom carried placards branding him the "Toxic Texan." The protests turned violent. Protestors threw stones and bottles, destroyed police cars, and threatened to storm the building where Bush was speaking.[16]

Returning to the United States and the Massachusetts Attorney General's Office in July 2001, Milkey surveyed the legal landscape before settling on a lawsuit to compel the EPA to regulate greenhouse gases as air pollutants under the Clean Air Act. He began laying the groundwork for an independent lawsuit by Massachusetts against the EPA, but upon learning about Mendelson's pending greenhouse gas petition before the EPA—and that the EPA was expected to formally deny the petition and claim greenhouse gases were not air pollutants—he pivoted to litigating the petition's denial. He contacted Mendelson to let him know of Massachusetts's interest in his case.

Milkey had no intention of being an incidental player in climate litigation. He wanted Massachusetts to be the lead state counsel in the case rather than to have California, Connecticut, New York, or Rhode Island seize the limelight. At the time these other states' attorneys general were California's Bill Lockyer, Connecticut's Richard Blumenthal (now senator), New York's Elliot Spitzer (subsequently governor), and Rhode Island's Sheldon Whitehouse (now senator). For these highly ambitious liberal politicians, each of whom relished the national spotlight, the chance to litigate the climate change issue had enormous political appeal.

To get out in front of them, Milkey decided he would quietly take advantage of an informal understanding that most states had that if more than one state brought a lawsuit against the federal government, the first to sue would take the lead role. Weeks before the EPA would ultimately deny Mendelson's petition, Milkey obtained advance approval from his boss, Attorney General Tom Reilly, to sue the EPA over the forthcoming denial. The governor at the time was Mitt Romney, but Reilly was a Democrat and Romney a Republican—and the attorney general was a completely independent actor. Romney played no role in the decision to bring the case.

The Clean Air Act provided that no lawsuit could be filed until thirty days after the EPA published its denial in the Federal Register. Milkey patiently waited, and one day after the EPA denied Mendelson's petition—still ten days before the agency would formally publish its decision and before any other state acted—he pounced, notifying other potentially interested states of Massachusetts's intent to file a lawsuit. No matter how many other states would ultimately decide to sue the EPA on the climate issue, Massachusetts (and Milkey) would now lead.[17] Even the California Attorney General's

Office, which continued to play a major role in the litigation and boasted a far larger number of expert environmental law attorneys, took a backseat to Milkey, though two attorneys from that office, Marc Melnick and Nicholas Stern, became especially important members of the team throughout the litigation. Mendelson welcomed Milkey on board, well aware of the strategic advantage of having a powerful state involved in the litigation.

Howard Fox and David Bookbinder

In addition to Mendelson, Doniger, and Milkey, two more attorneys from prominent national environmental organizations filled out the gang of five that coordinated the work of all the petitioners: Howard Fox from Earthjustice and David Bookbinder from the Sierra Club. A graduate of Yale College and NYU School of Law, Fox was a committed environmentalist. As a teenager he had attended the first Earth Day celebration on the National Mall in April 1970. And since the early 1980s he had worked for Earthjustice and its predecessor, the Sierra Club Legal Defense Fund. Now, in his late forties, his receding hairline accentuated his thin, tall frame and strongly angular face. Fox sported a dark paintbrush moustache, more in style in the 1970s than in the new millennium. Like Doniger, he was considered a leading expert on the Clean Air Act, an outstanding writer, and a gifted courtroom advocate (and accomplished ballroom dancer). But, unlike Doniger, Fox combined his sharp intellect with a soft-spoken, affable, modest, and polite manner.

The Sierra Club's Bookbinder rounded out the team. A decade younger, stockier, and a half-foot shorter than Doniger, Milkey, and Fox, with a fashionable short-cropped dark beard that thinned in its upper reaches, Bookbinder offered a natural bridge between the environmental organizations and the states. A summa cum laude graduate of Princeton University and the University of Chicago School of Law, his first environmental law job was working for

Milkey at the Massachusetts Attorney General's Office. This was after his Wall Street law firm supervisor advised Bookbinder to leave, noting—as Bookbinder himself later acknowledged—that he "didn't give a shit" about the kind of work the firm did—namely, representing large corporations fighting over money. Bookbinder went on from Massachusetts to work for a series of national environmental organizations before joining the Sierra Club, where he oversaw the organization's climate litigation and often worked closely with other groups like Doniger's NRDC. As a result, he had close personal and professional ties to both Doniger and Milkey.[18]

Bookbinder also enjoyed close ties to Mendelson. In December 2002, Bookbinder, alone, had supported Mendelson's filing of his lawsuit when other major national organizations were trying to scuttle it. He was impressed by Mendelson and thought his case had promise. At Bookbinder's urging, the Sierra Club had even joined the lawsuit over the strong initial opposition of the Sierra Club president. Sierra Club's support, both public and nonpublic, had prevented Mendelson from losing all of his funding.[19] Bookbinder was also the Sierra Club attorney who had contacted Milkey about Massachusetts joining Mendelson's case. In this way, Bookbinder often quietly worked behind the scenes, establishing multiple crisscrossing confidential relationships, in an effort to further climate policy.

Within this core group of five attorneys, Milkey sought from the outset to ensure that the states, led by Massachusetts, were the public face of the litigation. His reasons were in some respects very practical. Judges more readily see the states as legitimate spokespersons for the public interest than public interest organizations. States, after all, represent their citizens and are responsive to the will of the voters in elections. However well-intentioned nongovernmental public interest organizations may be, as representatives of the "public interest" they are self-appointed. That was why Milkey wanted a state—specifically, Massachusetts—to be the first-named petitioner in any lawsuit filed. His other, more personal reason: Milkey did not want to play second fiddle to anyone. He wanted to make sure he was the "counsel of record" for the petitioners challenging the EPA.

In a more harmonious world, NRDC's Doniger and Massachusetts's Milkey might have worked together seamlessly, and Bookbinder's bridging the two sets of petitioners might have proved not only tenable but highly constructive. After all, the two dominant personalities on the petitioners' side—Doniger and Milkey—had much in common. The two were from the Northeast, were of roughly the same age, and had similar elite academic pedigrees, extending even to master's degrees in urban planning on top of their law degrees. Most important, they also shared a strong commitment to compelling the EPA to address climate change.

But just as political foes can become unlikely allies, expected allies can also become foes, especially in the nation's capital, where extraordinary ability is routinely matched with dogged self-confidence. This seems to be what happened with Doniger and Milkey. By the time they reached the Supreme Court together almost two years later, in what should have been their shared mission to persuade the Court to rule in their favor, their professional and personal relations had long since been strained beyond the breaking point. They were not on speaking terms on the day of the argument, and have not spoken to one another since.

6

Weakness in Numbers

On October 23, 2003, approximately thirty parties filed petitions with the United States Court of Appeals for the District of Columbia Circuit, challenging the EPA's denial of Joe Mendelson's petition to regulate greenhouse gases from new motor vehicles.[1] The petitioners were twelve states—Massachusetts, California, Connecticut, Illinois, Maine, New Jersey, Minnesota, Oregon, New York, Rhode Island, Vermont, and Washington—as well as American Samoa, the Mariana Islands, Baltimore, New York City, and Washington, D.C. They also included several public interest organizations—now joining Mendelson's ICTA in the litigation were the Sierra Club, the Environmental Defense Fund, Earthjustice, the NRDC, and New England's Conservation Law Foundation.

Sierra Club's David Bookbinder hand-carried all of the petitions to the D.C. Circuit clerk's office for filing. Because all petitioners had agreed that the states should be lead petitioners and all the states had agreed that Massachusetts should be the lead state, Bookbinder handed the clerk the Massachusetts petition first and the clerk accordingly formally named the case *"Massachusetts v. EPA."*[2]

All thirty-plus petitioners filed their lawsuits in the same federal appellate court—the D.C. Circuit—rather than in any one of the hundreds of federal trial courts across the country. But not by choice. Most lawsuits against the federal government must first be filed in a federal trial court and reach a federal appeals court only if the party that loses decides to appeal the trial court's ruling. A plaintiff suing a federal agency generally has discretion to pick a federal trial court in a location most convenient to the plaintiff, which is typically where the plaintiff was injured or the allegedly unlawful action by the defendant took place.

Lawsuits brought under the Clean Air Act are different. Congress decided that lawsuits challenging certain kinds of actions taken by the EPA under the Clean Air Act not only must be brought in federal court in Washington, D.C., but must be brought in the first instance in the federal appellate court there (the D.C. Circuit Court), bypassing the trial court. The document filed with the appellate court is called a "petition" rather than a complaint, which is why Massachusetts and the others challenging the EPA were referred to as "petitioners" rather than as "plaintiffs," the term of art used to refer to parties filing complaints in trial courts. For that same reason, those opposing the petitioners are described as "respondents" and not as "defendants."

Congress had two reasons for bypassing trial courts for certain kinds of cases and having those cases decided by the same federal appellate court in D.C. First, it reasoned that the lawfulness of certain types of actions by a federal agency like the EPA should turn exclusively on whether the agency had acted reasonably based on the factual and legal arguments made to the agency at the time of its decision. Congress did not want trial judges to spend massive amounts of time developing their own trial records and then second-guessing the agency's fact-finding. Second, Congress decided that it was best to have the lawfulness of agency actions with nationwide impact, such as those taken by the EPA under the Clean Air Act, decided by a single federal appellate court in D.C. that could become expert in the kinds of administrative law issues that arise in such cases.

Knowing, however, that lawsuits like a challenge to the EPA's denial of Mendelson's petition routinely involve dozens, and often hundreds, of lawyers, the D.C. Circuit has established its own unique procedures for handling such cases. Rather than following the normal practice of allowing every party to file its own "brief"—the shorthand name for the document in which the party sets forth in writing its best legal arguments for why the court should rule in its favor—the D.C. Circuit sharply restricts both the number of briefs and the number of words that can be included in each side's legal brief.

For the judges on the D.C. Circuit, these rules are a godsend, eliminating thousands of pages of legal briefs that would be making mostly duplicative arguments. This gift to the judges, however, is a nightmare for the hundreds of lawyers involved in representing multiple petitioners and respondents in any one case. Even if many of the lawyers on either side agree about many things, they still represent (and are paid by) different clients with different priorities. Additionally, the lawyers invariably have very different views on what are the best and most effective legal arguments to make. Reaching a consensus in a single, unified brief filed on behalf of either all or a large subset of petitioners or respondents is enormously difficult. But under the D.C. Circuit rules, there is no other way to proceed.

Consistent with those rules, the D.C. Circuit in *Massachusetts v. EPA* established a briefing schedule for both petitioners and respondents. The petitioners had to file one opening brief of no more than 18,000 words (approximately 72 pages) by June 22, 2004. The court, in turn, gave the EPA until October 12, 2004, to file its own brief, also of no more than 18,000 words, in response to the arguments made in the petitioners' opening brief.

The petitioners and the EPA were not the only parties to the lawsuit. There were also several states and private industry respondents that supported the EPA. The court permitted these two separate groups of respondents to file by the October 12 deadline their own short briefs of no more than 4,375 words (approximately 17 pages). And, finally, the petitioners were allowed to file a joint "reply brief" by December 17, 2004—giving them a chance to answer the arguments made by all the respondents.

In choosing to file a single joint brief, the petitioners made a strategic choice. Even under its restrictive rules designed to discourage multiple briefs, the appellate court would likely have allowed petitioners to file two separate briefs—one from the states and one from the public interest organizations—with each brief allotted half the total word count. The petitioners, however, agreed to submit instead a single, longer, and what they hoped would be more effective joint brief. They assumed that their shared litigation objectives would make it possible to reach the necessary consensus regarding the content of their brief. This promised the public interest organizations some influence over the content of the states' legal arguments, but it also meant they would lose their independent voice in the brief filed with the appellate court.

Forced to coordinate their arguments, the petitioners divided their joint brief into as many as ten different parts and assigned as many as ten different attorneys to assist in the initial drafting of each of those parts.[3] Their initial drafts, along with every subsequent draft, would then be circulated among the dozens of attorneys representing all of the petitioners for comment and proposed edits until the final brief was ready for filing by June 22.

Any initial hope that they could all work together and easily reach consensus on the brief's contents was quickly dashed. The petitioners' fragmented process bordered at times on the disastrous, seriously undermining the quality of the brief they jointly filed in June. First and subsequent drafts were subjected to dozens of reviewers with different ideas, priorities, and writing styles. Rather than gradually moving toward agreement, petitioners with sharply contrasting perspectives dug in ever deeper, growing more insistent and less willing to compromise.

What began as polite disagreements during the first few rounds of editing became sharply worded critiques, sometimes leveled in highly personal terms. Instead of modest, discrete edits on a single version of an argument, competing attorneys circulated complete re-writes of entire sections of briefs. And instead of fully transparent email exchanges between all the lawyers involved, there were clandestine communications between subgroups.

Most often the divide centered on differences in view between the states and the environmental groups. Even more directly, it centered on differences between Milkey and Doniger. Lacking any formal hierarchy to resolve such disputes, communication, trust, and respect broke down as the June filing deadline rapidly approached.

Two parts of the petitioners' brief were the greatest casualties: the argument that greenhouse gases are air pollutants (for which Milkey served as the lead drafter) and the argument that the EPA had further erred in not deciding whether new motor vehicle emissions of greenhouse gases endanger public health and welfare (for which Fox served as the lead writer). Ironically, these two parts of the brief suffered disproportionately from the chaotic briefing process for nearly opposite reasons: Milkey's because the available legal arguments were so strong, and Fox's because the available legal arguments were comparatively weak.

Greenhouse Gases as Air Pollutants

Jim Milkey's portion of the brief, addressing why greenhouse gases constituted air pollutants, suffered from the perverse problem that the core argument was so strong that there were numerous ways it could potentially be pitched. As first outlined in Mendelson's draft petition to the EPA five years earlier, the central point of the legal argument bordered on the irrefutable: the Clean Air Act defines an "air pollutant" as a chemical compound emitted into the "ambient air"—which refers to the air in the immediately surrounding atmosphere—and greenhouse gases easily meet that definition.[4] Carbon dioxide—the most plentiful of greenhouse gases—methane, nitrous oxides, and all the more complicated fluorinated compounds that contribute to the "greenhouse effect," and therefore climate change, are all "chemical compounds" that are "emitted into the ambient air." There was no room for debate. The only remaining legal question was whether the EPA had authority to adopt an interpretation of "air pollutant" in contradiction to the plain meaning of the Clean Air Act's definition of that term.

In 1984 in one of its most frequently cited and influential cases of the past forty years, *Chevron v. Natural Resources Defense Council, Inc.,* the Supreme Court had established certain rules that courts must follow in deciding when a federal agency has adopted a valid reading of a federal statute that the agency is responsible for administering. According to *Chevron,* a court should defer to a federal agency's interpretation of statutory language only when the meaning of that language was ambiguous.[5] When the language had a plain meaning, a court should reject any agency reading of the law contradicted by that plain meaning. Milkey accordingly had a strong argument that *Chevron* meant that the EPA's conclusion that greenhouse gases are not air pollutants should be squarely rejected because it was contrary to the Clean Air Act's clear statutory definition of "air pollutant."

In denying Mendelson's petition, the EPA had proffered a large number of peripheral reasons the D.C. Circuit should, notwithstanding *Chevron,* uphold the agency's view that greenhouse gases were not air pollutants. Most of the agency's rationale focused on the fact that there was little, if any, suggestion that Congress had contemplated climate change when it passed the law in 1970 that included the definition of "air pollutant." The EPA argued, in effect, that the Clean Air Act's broad definition of air pollutant should not be understood to include greenhouse gases unless and until Congress expressly considered, debated, and decided how the climate issue should be addressed. The breadth of the statutory language was not, standing alone, sufficient to answer the legal question.

Were it not for the Supreme Court's decision three years earlier in *U.S. Food and Drug Administration v. Brown & Williamson Tobacco Corp.*, the EPA's argument would have seemed wholly implausible. But in *Brown & Williamson*, the Court had ruled that the Food and Drug Administration, a federal agency within the Department of Health and Human Services, lacked authority to regulate tobacco even though the plain meaning of the Food and Drug Act seemed capacious enough to include tobacco as a "drug." Given the enormous political and social consequences of regulating tobacco as a drug, the Court reasoned that Congress should not be understood to have authorized this without evidence that it had specifically intended such a result.[6]

Petitioners had a series of potentially powerful ways to distinguish regulating tobacco as a "drug" from regulating greenhouse gases as a "pollutant," but how best to make that distinction proved highly divisive. They disagreed on which of several possible arguments to make, or not make, what order to make them in, and how to phrase them.

Milkey circulated a first draft of the argument to the entire team in early March, more than three and a half months before the brief was due.[7] But several weeks later the petitioners seemed in many respects farther from consensus. Milkey's many co-counsel had inundated him with comments and edits, big and small. Some called for major stylistic changes, others for wholesale rejections of specific arguments. Many of the comments were in conflict with each other. There was no way for Milkey to craft a revised draft that would satisfy everyone.[8]

By mid-May, with no clear notice from Milkey about when the revised draft would appear, some of Milkey's co-counsel expressed concern. The full brief needed to be filed in late June, in little more than five weeks. An attorney from the California Attorney General's Office wrote Milkey on May 12, "What's up?," then added, with deliberate understatement that did not greatly mask rising concerns within the litigation team, "I'm a little concerned that we're behind schedule."[9]

Milkey finally emailed his revised draft to co-counsel on May 24.[10] As with the first draft, reactions were mixed. One voice, however, was most vocal in its criticism: Doniger's.

Rather than edit Milkey's draft, Doniger circulated a complete re-write of a significant part of the argument.[11] Among other things, Doniger faulted Milkey for not doing more to address a weakness in their argument that Doniger anticipated the EPA would raise in its brief—the assertion that the Supreme Court's prior decision in *Brown & Williamson* that tobacco could not be regulated by the FDA meant that greenhouse gases could not be regulated by the EPA under the Clean Air Act. Doniger believed that it would be far better to grab the bull by the horns in the opening brief—and address their toughest argument more fully—than to leave themselves open to attack. But Doniger's approach was rebuffed by another in the legion of lawyers reviewing the now-competing drafts, who characterized Doniger's proposed argument as "the equivalent of hanging a sign on your back saying 'please kick me.'"[12]

By early June, any suggestion of an orderly process for drafting Milkey's portion of the brief had broken down. Doniger continued to circulate his own versions of the argument rather than adhere to their prior norm of editing Milkey's draft. When Milkey sought to reconcile his and Doniger's versions in yet another revised draft, Doniger rebuffed the effort. On June 9, Doniger sent Bookbinder a strongly worded 1:00 a.m. email, asking Bookbinder to share his draft with the full team: "This is not something that can be solved by editing," he wrote. "The problem is that [Milkey's] responses are fundamentally unconvincing—worse, helpful to the EPA—and seriously damaging to our overall case."[13]

With only eight days before the filing deadline, the conflict hardened with no signs of abatement. On June 14, Milkey wrote all petitioners' counsel with yet another draft. Milkey's accompanying email revealed his deepening frustration. He complained about being repeatedly asked to make changes he had previously declined to accept. He sharply criticized Doniger for describing as "gratuitous" a portion of the argument that Milkey thought "drives home an important point." In response to another one of Doniger's repeated suggestions, Milkey stated that he had declined to accept it the first time it was offered "because the subtlety of the reference escaped me" but had "knowingly declined" it the second time.[14]

Privately, Milkey was angry. Doniger, he thought, was ambushing his work. Especially infuriating was Doniger's habit of circulating competing drafts rather than offering edits of Milkey's draft. Doniger, for his part, felt he had no choice due to flaws in Milkey's reasoning. He could not agree to what he considered subpar work.[15]

As the June 22 filing deadline loomed, Milkey decided to surrender. He did so not because he agreed with Doniger on the substance. Nor was he any less angry about what he perceived as Doniger's bullying. Milkey stepped aside only because he knew how important it was that they reach closure on the argument.[16] They were almost out of time and still needed to proofread carefully the entire brief to make sure all the hundreds of references in support of their arguments—judicial precedent, statutory provisions, agency regulations, scholarly publications—were entirely accurate. Judges are exacting. A brief with any mistakes, whether typos, grammatical errors, or incorrectly cited legal authority, lacks credibility and persuasiveness.

Doniger had no regrets about any possible heavy-handedness on his part during the editing process. It was precisely because he did not trust others to file the best possible brief in a case of such exceptional importance that he had argued in favor of a joint brief rather than having the states and the environmental groups file two separate briefs. In early meetings, he said he favored a joint brief because one brief would be "more effective" and would allow petitioners to "work nuance to consensus."[17] But in reality Doniger favored a single brief because he wanted to "have more control" over the brief that was being filed, including the "opportunity to veto" arguments that he considered substandard.[18]

To outsiders, the differences between Doniger's and Milkey's approaches were not nearly as great or as important as either perceived them to be. There are always many ways of wording any legal argument. Choices need to be made about which wording is clearer and more persuasive, and which possible weaknesses in one's argument should be addressed in anticipation of the other side's brief. In most instances, the actual stakes are much lower than they are perceived to be at the time, and the inevitable disagreements—arising out of good-faith differences among petitioners—are resolved with a few bruised feelings that are quickly forgotten days after the brief's filing. In *Massachusetts,* however, because of the sheer number of lawyers involved in overseeing the drafting process, the competing personalities of the key players, and the overwhelming significance of the climate issue, the intensity of the dispute did not diminish. It remained a festering wound.

The Backup Argument

While Milkey was working on what should have been the easy argument that greenhouse gases are clearly air pollutants, Earthjustice's Howard Fox was drafting the portion of the brief that responded to the EPA's backup argument that the agency could lawfully deny Mendelson's petition even if the court concluded that greenhouse gases were air pollutants. Fox was hugely admired by all and never rubbed anyone the wrong way. But here too, the petitioners' lawyers sharply disagreed about how best to craft the argument, though in far less personal terms than they had with Milkey's draft.

Fox faced a significant threshold problem in crafting an effective response. The EPA's decision was written in such a confusing manner that it was hard to say precisely what its reasons were and, therefore, was difficult to respond to. The EPA's backup argument could be read in either of two ways: (1) as concluding that emissions of greenhouse gases do not endanger public health and welfare; or (2) as asserting the EPA's right not to decide yet whether greenhouse gas emissions endanger public health and welfare. The EPA's political appointees had insisted on language that suggested the former because it closed the door on climate regulation even more firmly. But the agency's career employees favored the latter argument precisely because it preserved the EPA's options—a favorite approach of career employees regardless of the administration in power. The two groups (the political appointees and the career public servants) had compromised by including language in the EPA's decision that reflected both views.

If the agency had simply concluded that greenhouse gases do not endanger public health and welfare, Fox would have had a great argument that the EPA had acted unlawfully. By 2003 the climate science contradicting such a conclusion was massive. But if the EPA's claim was instead that it had the discretion to decline to decide the endangerment issue right now, Fox would have a much harder legal argument to make. Courts routinely give federal agencies lots of discretion in deciding when to make decisions.

Fox had a dilemma. He could argue that the EPA had based its denial of Mendelson's petition on a finding that greenhouse gases did not endanger public health and welfare, the easier argument for petitioners to defeat. Or he could acknowledge the alternative way to read the EPA's argument and take on the far tougher argument in the opening brief rather than leave it only for possible cursory discussion in the sharply limited pages available in their final reply brief, to be filed after respondents filed their own briefs. Petitioners' lawyers were sharply conflicted on how best to proceed. They disagreed about what the EPA's actual reasoning was. They disagreed about whether they should take advantage of the confusing language in the EPA's decision by focusing only on the agency's weaker argument. And they disagreed about how best to respond to the EPA's stronger argument if they were to address it in their brief.[19]

Fox circulated the initial draft of his argument on March 16 for review by all the co-counsel.[20] The draft principally characterized the EPA as having concluded that greenhouse gases do not endanger public health and welfare and, on that basis, forcefully challenged the validity of any such conclusion in light of climate science. The draft did not deal in a focused way with the possibility that the EPA had simply declined to decide the endangerment issue.

The negative response from some quarters was politely worded but pointed and immediate.[21] Two attorneys from the California Attorney General's Office were especially skeptical. One faulted the draft for focusing on a false target and for failing to recognize that "[the] EPA clearly claims that it has not yet exercised its judgment, one way or the other" and that it does not need to do so.[22] Another, while characterizing Fox's draft as a "great start," suggested that the draft should spend more words on addressing what was in fact the petitioners' hardest issue: why the EPA had a mandatory duty to make an endangerment determination, given that its decision could be fairly read as declining to make one.[23]

Still others on the legal team pushed back in support of Fox's approach.[24] The back and forth prompted one of the draft's critics to joke that others thought he "must be smokin' crack."[25] The internal debate persisted for months, well into mid-May, prompting a formal conference call to discuss what Milkey characterized as "a lively e-mail exchange that revealed many different perspectives."[26]

Milkey offered a different way to pitch the argument for this last part of their brief. He speculated that they could exploit the fact that "[the] EPA titles its discussion of why it chooses not to regulate greenhouse gases: 'Different Policy Approach.' We could use this to emphasize that the EPA in the end simply disagrees with the policy choices enunciated in the Clean Air Act." There "is promise here," Milkey thought, to argue that the EPA had acted unreasonably by relying on an impermissible ground—"policy choices" different from those Congress made—in deciding not to decide the endangerment issue. The EPA, in short, could not lawfully reject Congress's policy choices in favor of its own.[27]

In late May, with the due date approaching, one member of the team quietly and privately brought in a truly extraordinary outside expert to review their draft brief. The reviewer was former D.C. Circuit chief judge Patricia Wald, who had retired from the bench only a few years before. Wald was justifiably considered a giant in the legal profession and, in light of her years on the D.C. Circuit, uniquely able to advise the *Massachusetts* petitioners on how best to pitch their arguments to her former colleagues on that court. The judge offered her review on a strictly pro bono, confidential basis as a personal favor to one of the many attorneys involved in the litigation who was a close family friend. Her affiliation with the case has never before been revealed and only a very few on the team were even aware of Judge Wald's involvement.[28]

Judge Wald's advice was sobering. She largely agreed with those who said their argument on this second issue was weak. She wrote that it was hard to articulate why the EPA's decision—in effect to decide not to decide the endangerment issue—was actually "unreasonable." This was, she bluntly concluded, their "toughest argument to make." After pointing out that difficulty, and strongly suggesting that the petitioners would have difficulty prevailing, the best Judge Wald could muster was a closing "Good luck!"[29]

The brief that was submitted did not heed the advice of the critics within the petitioners' litigation team. Rather than single out for distinct, focused attention what its critics had identified as the EPA's stronger argument, the brief chose to take advantage of the truly confusing nature of the EPA's published analysis by conflating its stronger argument with its weaker argument, hoping to undermine the former by its association with the latter.

Writing around the hardest legal questions was a risky move. Unless the opposing counsel are incompetent or the judges unprepared, the proverbial chickens will almost always come home to roost. Opposing counsel will later clarify their presentation at oral argument by identifying their best argument, or the judges will do that for them. And, when they do, one's ability to answer the hardest question effectively the second time around will be severely limited—both because the advocate's initial lack of candor undermines his credibility before the court and because a judge will very likely raise the issue at oral argument, which is the advocate's weakest platform for addressing issues for the first time. In oral argument the judge controls everything, from the questions asked to the time allowed for response.

The petitioners had in effect gambled that the D.C. Circuit judges would fail to identify the EPA's stronger argument on their own. That was exceedingly unlikely to happen, as Judge Wald had effectively warned. And it didn't.

7

Three Judges

On April 8, 2005, when Jim Milkey rose before the D.C. Circuit to present oral argument in *Massachusetts v. EPA*, Courtroom No. 31 was packed. It was the lawyer's equivalent of a shopping mall on Black Friday, but instead of informally clad frantic bargain hunters with overexcited children in tow, the room was full of measured attorneys in gray and dark-blue suits. The court clerk had reserved a total of sixty seats just for the lawyers in the case: thirty for the *Massachusetts* petitioners and another thirty for the lawyers representing the EPA and parties supporting the agency. And that was a mere fraction of the total number of attorneys who had filed briefs in the case and hoped to attend the argument. As a result, there were relatively few seats available for interested members of the public. It was standing room only, with spectators lined up against the courtroom's back wall. The overflow crowd was sent to another courtroom, where the audio of the argument was piped in.

Yet even though almost all of the attorneys and spectators in the courtroom and the overflow room that morning were there because of the *Massachusetts* case, it was not the highest-profile case being heard that morning in Courtroom No. 31 of the E. Barrett Prettyman

United States Courthouse, which is located on Pennsylvania Avenue about midway between the U.S. Capitol Building and the White House. That honor went to *Pro Football v. Susan Harjo,* argued immediately beforehand, in which the judges reviewed the decision of the U.S. Patent Office to cancel the trademark of the National Football League's Washington Redskins on the ground that the "Redskins" name was disparaging to Native Americans.[1] The big news stories of the day concerned the fate of the Redskins, not the fate of the planet threatened by climate change. It was not until later that year, in early September, that the American public would begin to think hard about the harsh reality of climate change in the aftermath of the devastation caused by Hurricane Katrina to New Orleans.

The same three federal judges heard both cases. All three were highly accomplished, seasoned jurists. They all wore identical black robes intended to underscore their shared commitment to the neutrality of the administration of justice. And they were understandably proud of their court's reputation for reaching consensus. Yet the judges inevitably brought to their work differences in experiences and outlook that could cause them to reach very different conclusions in some cases. *Massachusetts* was just such a case.

Although the D.C. Circuit is technically just one of thirteen federal courts of appeals, it is often touted as the most prestigious both because of its location in the nation's capital and because it hears many of the biggest cases relating to the authority of Congress and the president. The D.C. Circuit is also seen as a stepping stone to the Supreme Court. At the time of the *Massachusetts* argument, four of the Justices then on the Supreme Court—Chief Justice John G. Roberts Jr., and Associate Justices Antonin Scalia, Clarence Thomas, and Ruth Bader Ginsburg—had first been D.C. Circuit judges.

In October 2003, when the *Massachusetts* petitioners filed their lawsuit in the D.C. Circuit, nine judges sat on that court along with two retired senior judges who were also available to hear and decide cases.[2] But not all D.C. Circuit regular and senior judges vote in every case. Like all federal courts of appeals, the D.C. Circuit establishes, by random assignment, three-judge panels to read the briefs, hear the oral argument, and draft the opinion for the entire court. Only very rarely do all the judges sit together *(en banc)* to decide a case.

In the D.C. Circuit, the parties don't learn the identity of the three judges hearing their case until the court issues a formal order setting the date for argument, which may or may not be before the briefs are filed. In *Massachusetts,* the parties learned on May 24, 2004, only a few weeks before the petitioners' brief was due, both the names of the three judges on their panel and the oral argument date: April 8, 2005, almost a year later. (In some other federal courts of appeals, the parties do not learn the names of the judges deciding their case until the morning of the oral argument.)

Lawyers are under no illusion that it makes no difference which three judges are randomly assigned to their case. It can have enormous consequences. To be sure, the D.C. Circuit judges are all among the most talented federal judges in the country. But even within that rarified group, there are marked differences in ability and temperament, and there can be significant ideological differences.

In the vast majority of cases, such ideological differences will not affect how a judge votes. The clarity of the relevant law, the three judges' shared commitment to adhering to the precedent established by their court's own prior rulings, and neutral principles of judicial decision-making usually lead them all to support the same outcome. That is why the D.C. Circuit has few dissents, even though the judges on that court were appointed by presidents of sharply different political leanings.[3] But in some cases, the panel's makeup can make a big difference in the outcome—especially when the applicable law is less clear, or when the kind of legal issue raised in a specific case taps into differences in how individual judges perceive their role as judges.

Some judges are committed to the proposition that the text of a statute or a constitutional provision is the best, or should even be the only, guide in interpreting its meaning, while others are more willing to look at other indices of underlying intent. Some judges grant great deference to the fact-finding and policy judgments of other parts of the government—executive branch agencies, congressional committees, and state and local governments—while others do not. These differences in judicial outlook can determine the outcome in individual cases.

In *Massachusetts*, both sides understood that the identity of the three randomly assigned judges could make a big difference in the outcome of the case. After all, the case had strong political overtones, headlined by state attorneys general affiliated with the Democratic Party that were challenging the high-profile actions of a Republican president. The *Massachusetts* petitioners also understood that the kind of judicial relief for which they were asking—second-guessing the executive branch on the need for tougher environmental regulation—might strike some conservative judges as smacking of the very type of "judicial activism" they adamantly opposed.[4]

That is why the EPA was heartened and the *Massachusetts* petitioners concerned when they learned that their three judges would be David Sentelle, Raymond Randolph, and David Tatel. Sentelle and Randolph had been appointed by Republican presidents (Ronald Reagan and George H. W. Bush, respectively) and Tatel by a Democrat (Bill Clinton)—but the identity of the nominating president is an unduly simplistic and frequently misleading predictor of judicial behavior. The parties' reactions were based on the judges' contrasting professional backgrounds and established track records in decided cases on the D.C. Circuit. All three were about the same age (sixty) and none was new to the bench. Sentelle and Randolph had been on the D.C. Circuit since 1987 and 1990, and Tatel had joined them in 1994. A lot was known about all three.

Judges David Sentelle and Raymond Randolph

During their many years on the bench together, Judges Sentelle and Randolph had generally proven skeptical of lawsuits in which environmentalists had asked judges to compel a federal agency to impose more demanding regulations on business than the agency deemed appropriate. They had both come of age on the court at a time when what it meant to be a "conservative" judge was to decline to be the kind of "judicial activist" Presidents Reagan and George H. W. Bush had campaigned against in running for the White House.[5] They believed in judicial restraint—a shorthand for judges not second-guessing decisions by other branches of government.

Judge Sentelle was born and raised in Canton, North Carolina, a town of about four thousand, where his father was a mill worker. He had gone to the University of North Carolina for both college and law school and had been politically active with the Republican Party on campus, serving as president of the Young Republicans and chair of the conservative group Young Americans for Freedom. He had worked as a local federal prosecutor, state court judge, and in private practice and had served as an alternate delegate to the 1984 Republican National Convention before being appointed by President Ronald Reagan to serve as a federal judge in North Carolina in 1985.[6]

Elevated to the D.C. Circuit two years later by President Reagan, Judge Sentelle was known as a demanding but charming judge with a knack for storytelling. Boasting a loud, deep Southern drawl and a penchant for wearing a cowboy hat and boots and driving a pickup truck, Sentelle had a thick mane of hair, and was well liked and respected by his colleagues on the bench across the political spectrum. He enjoyed a deserved reputation for adhering to precedent established by the court's past rulings, even if he might disagree with those rulings.[7] Apart from his involvement in the selection of Ken Starr to serve as special counsel in the investigation of President Bill Clinton, which sparked controversy because of his close ties to the Republican Party, Sentelle was rarely in the national limelight.[8]

Judge Randolph shared Judge Sentelle's conservative political outlook and also was raised in a rural area (in his case in New Jersey) by parents of modest means.[9] But Randolph and Sentelle were two very different people, as anyone could sense immediately from their appearance, demeanor, and self-presentation. While Sentelle was tall, wide, and built like a football player, Randolph was short, compact, and built like a wrestler, and in fact he had been a member of his college's varsity wrestling squad. While Sentelle had gone to public schools and universities, Randolph had attended both a private college (Drexel University) for his undergraduate studies and a private law school (University of Pennsylvania School of Law).[10] Finally, while Sentelle was charming and engaging, and admired by all of his colleagues, Randolph was intense, serious, and at times downright unpleasant.

At Drexel, Randolph had first studied business administration, but he has since acknowledged in personal interviews that he had "absolutely hated" the subject and "dropped out after three months." Unsure what he wanted to do, Randolph next studied engineering; but when he "saw what some of the engineers were doing at the General Motors Plant," he quickly concluded that he wouldn't be "happy . . . doing that sort of thing," so he pivoted to economics and, after reading a history of Justice Felix Frankfurter, decided on law school. He was the first ever in his family to study law.[11]

It was not until he was in law school at Penn that Randolph began to identify more with conservative Republican politics, despite his family's long-standing roots in the Democratic Party. The shift was prompted by his disagreement with the liberal rulings of the Supreme Court under Chief Justice Earl Warren. After graduating summa cum laude and first in his law school class from Penn, Randolph spent most of his career in private practice in Washington, D.C., before becoming a federal appellate judge.[12]

Earlier in his career, however, he had served two stints at the Solicitor General's Office of the U.S. Department of Justice, where he had argued cases in the Supreme Court on behalf of the United States. In his second tour in that office, he served as deputy to Solicitor General Robert Bork.[13] A Yale law professor, Bork became a hero of legal conservatives because he criticized "activist judges" who effectively legislated from the bench by failing to adhere to the Framers' "original understanding" of the U.S. Constitution. This prompted both President Ronald Reagan to nominate Bork to the Supreme Court and the Democratic-controlled Senate to reject the nomination.[14] Randolph's prominent support of Bork's nomination likely played a role in President George H. W. Bush's decision to appoint him a few years later to the D.C. Circuit, where Bork himself had been a judge. Once on the appellate bench, Randolph had been a reliably conservative judge, but he authored no high-profile opinions and rarely generated any publicity.

Judge David Tatel

Judge Tatel was the *Massachusetts* petitioners' dream pick for the three-judge panel. Although Tatel's immediate job before joining the D.C. Circuit in 1994 was with a prestigious D.C. law firm, he had devoted much of his earlier career to championing social justice.[15] A native of Washington, D.C., and a graduate of the University of Chicago, Tatel had been a highly renowned civil rights lawyer who assisted in the desegregation of public schools across the United States in the 1970s, 1980s, and 1990s.[16] He had served as the founding executive director of the Chicago chapter of the Lawyers' Committee for Civil Rights Under Law, as director of the National Lawyers' Committee for Civil Rights Under Law in D.C., as the first general counsel of the Legal Services Corporation, and as head of the Office of Civil Rights in the Department of Health, Education, and Welfare during the administration of President Jimmy Carter.[17]

One of the most prominent photographs on the wall in Tatel's judicial chambers was a picture of Tatel with Justice Thurgood Marshall, a civil rights icon.

On the court, Tatel had quickly established a national reputation as a brilliant, energetic jurist. Lithe and athletic, he had run multiple marathons and was an avid skier (both snow and water), hiker, and windsurfer. He went swimming three times a week before work.[18] Like Judge Sentelle, Tatel was well liked and highly respected by his colleagues. Although their personal politics were light years apart, each admired the other greatly, and they had developed a strong professional and personal bond over their years on the bench together.[19]

In the months preceding the 2000 presidential election, Tatel had been touted as a likely nominee to the Supreme Court in the event that Al Gore won that election.[20] When George W. Bush was instead elected in 2000—and reelected in 2004—Tatel's pathway to the Court effectively closed due to the modern penchant of presidents to nominate to the Court people less than sixty years of age. Tatel would be sixty-six in 2008.

On the bench, Tatel had shown a particular interest in environmental law issues and a receptiveness to lawsuits filed by environmentalists against the government. Steeped in the history of civil rights, Judge Tatel naturally saw the judicial role as critical to ensuring the fair and full enforcement of the nation's laws. He was well aware, from his experience with civil rights law, of the tendency, absent a judicial check, of both government agencies and the private sector to fail to comply with the strict requirements of the law.[21]

Environmentalists perceived that Judge Tatel was aware of that same tendency in environmental law and that he appreciated the heightened need for judicial scrutiny to guard against governmental shortfalls in environmental enforcement. Tatel had not voted reflexively in favor of environmentalists—he was a stickler for rigorous legal analysis—but the *Massachusetts* petitioners were confident of the strength of their legal arguments and saw him as a potential champion of their position on the bench.

One potentially sympathetic judge out of three, though, did not a majority make. Upon learning the names of the three judges on their panel, the *Massachusetts* petitioners understood that their odds of prevailing had just gone down. Considerably.

8

Completely Confused

As the petitioners, the thirty parties challenging the EPA's denial of Joe Mendelson's petition were first to the lectern for the oral argument. The court had allotted twenty minutes to each side, and the *Massachusetts* petitioners had decided to have both Milkey and Howard Fox present oral argument. They further decided that Milkey would have fourteen of those minutes to address the legal issue for which he had served as the principal brief writer: whether the EPA's decision was invalid because the agency had unlawfully concluded that greenhouse gases are not "air pollutants" within the meaning of the Clean Air Act. After Milkey, Fox would have the remaining six minutes to address the separate legal issue of whether the EPA's denial of Mendelson's petition could be sustained on its backup contention that the agency could lawfully decide, at least for now, not to regulate greenhouse gases from motor vehicles even if those gases were air pollutants.[1]

There was a fatal flaw in the petitioners' plan. They had violated the most important rule of oral advocacy. A lawyer presenting oral argument in any court has one job and one job only: answer the judge's questions. The judges are in control of the questions, their order, and how much time they want to spend on different legal issues. The advocate does not get to pick the questions, cannot decline to answer them without dire consequences, or get to decide how much time to spend answering any one question. The effective oral advocate must be nimble and be ready to answer whatever question the judges ask, when asked and within the short time period available before the same judge or a different judge abruptly interrupts to ask a new question.

As would very soon be made all too clear, the petitioners' decision to divide their argument into two parts undermined their ability to be responsive to the judges' questions. The only saving grace was that the attorney representing the EPA fared even worse when it was his turn before the lectern. April 8, 2005, was not a good day for oral advocacy.

Wasted

Within seconds of beginning his argument, Milkey faced the consequences of this serious misstep. The petitioners may have wanted to spend fourteen minutes of their time talking about whether greenhouse gases were air pollutants, but the three judges did not. In fact, the judges showed no interest in discussing that issue at all. They only wanted to ask questions about the issue that Fox was supposed to discuss. But Milkey was neither prepared for these questions, nor had he really been authorized by his many co-counsel to answer them.

So when Judge Sentelle interrupted Milkey at the outset of his argument and asked him to address the EPA's "fallback position," which stipulated that even if the EPA is authorized to regulate greenhouse gases, "they're not required to," Milkey could only weakly respond that his co-counsel Fox "is going to be addressing that" next.[2] And when Milkey responded to a subsequent question by Judge Randolph by similarly saying Fox "will address that" too, Sentelle was visibly angered.[3]

Any pretense of Southern charm was gone. Judge Sentelle accused Milkey of repeatedly dodging the tough questions by saying that his co-counsel was going to address them later. Because none of the judges had any interest in the only issue that Milkey was prepared to discuss, Milkey's fourteen minutes—more than two-thirds of petitioners' total allotted time—were largely wasted. Worse, petitioners had alienated the bench. Once that happens, few advocates can recover. Milkey didn't, though at the time he was too immersed in the moment to appreciate how poorly his argument had gone.

Fox was next to stand before the lectern and face the judges. Unlike Milkey, who had not previously argued before the D.C. Circuit and had argued mostly in the Massachusetts Supreme Judicial Court, Fox was well known to the D.C. Circuit judges. He had a deserved reputation for being an expert on the Clean Air Act and for his clear speaking style, and he was very highly regarded. Fox had previously argued ten Clean Air Act cases before the D.C. Circuit, many before the same judges hearing *Massachusetts*. Judges looked forward to his arguments.[4] But like Milkey before him, Fox fell short that morning.

He paid the price for the weaknesses in the written brief that he and his co-counsel had filed, and for choosing to take advantage of the confusing way in which the EPA had explained its reasons for denying Mendelson's petition. They had conflated rather than separately addressed the stronger and weaker versions of the EPA's argument.[5]

It proved an unwise strategy. Federal appellate court judges are smart lawyers, and they had no problem identifying the EPA's stronger argument on their own. Finding no clear, direct answer in the petitioners' brief to the EPA's argument, the judges repeatedly demanded that answer at the oral argument, first from Milkey and then from Fox. Having failed to use their written brief to craft an effective answer, Fox was limited to providing the best explanation he could come up with on his feet in the few seconds available between one hostile question and the next.

Judge Sentelle wasted no time in asking Fox the same question he had unsuccessfully put to Milkey earlier. He challenged Fox almost as soon as he stood up to explain how he could square the petitioners' contention that the EPA had a mandatory duty to regulate greenhouse gases with the statutory language. As Sentelle explained, that language required the EPA to regulate emissions only after first exercising its "judgment" to conclude that greenhouse gas emissions endangered public health and welfare. Given that the EPA had not yet come to such a conclusion, Sentelle continued, what was the basis of petitioners' argument that the EPA's mandatory duty to regulate greenhouse gas emissions had been triggered?

Fox, repeatedly interrupted as he spoke, was never able to clearly explain the petitioners' legal argument. Nor, given the few minutes that had been allotted to his portion of the argument, did he have much of a chance to do so. Rather than answer Sentelle's question and clearly state whether the petitioners were contending that the EPA had a mandatory duty to regulate greenhouse gases, Fox sought to sidestep the issue by explaining that the petitioners were not asking the court to order the EPA to regulate greenhouse gases. But then he added that the EPA had misinterpreted the meaning of the word "endangerment" in the Clean Air Act, leaving the judges with the misimpression that the petitioners might in fact be arguing in favor of such a mandatory duty. In all events, neither of Fox's responses clearly answered Judge Sentelle's question. Worse, they were potentially contradictory.[6]

By the close of Fox's argument, the only thing that was clear was that neither Sentelle nor Randolph understood Fox's argument. Even a highly skilled oral advocate like Fox could not overcome a weak brief.

A Judicial Buzz-Saw

When Fox sat down, Jeffrey Bossert Clark, deputy assistant attorney general of the Justice Department's Environment and Natural

Resources Division, presented his oral argument on behalf of the EPA. The Justice Department is celebrated for its litigation expertise. That is a major reason federal agencies like the EPA cannot represent themselves in litigation and are represented in court by Justice Department lawyers. Prior to litigation, the EPA is in charge of its own decisions, but once a case goes to the federal court, the EPA becomes a client and the Justice Department has the final word on all litigation decisions, including the substance of the arguments to be made.

Clark's appearance before the court sent a clear message to the three judges sitting on the bench: the Bush administration was treating this as a high-profile *political* case. Clark was not standing before them because of his litigation expertise, let alone his expertise arguing cases before the D.C. Circuit. He was presenting the oral argument because he was one of a half-dozen "political appointees" among the five hundred career attorneys in the Justice Department's Environment and Natural Resources Division.

Thirty-eight years old and a graduate of Harvard College and the Georgetown University Law Center, Clark was a still relatively young attorney with a promising political trajectory. After graduating with high honors from Georgetown, he had clerked for a distinguished federal appellate judge before joining the D.C. law office of Kirkland & Ellis, a prestigious Chicago-based law firm known for its strong environmental law practice and its aggressive litigation style. He had been at Kirkland for five years when he had jumped at the opportunity for a political appointment in the Bush Justice Department's Environment Division in early 2001.

But unlike career Environment Division appellate attorneys, who have argued dozens and sometimes several hundred cases in the federal courts of appeals, including the D.C. Circuit, Clark was an inexperienced oral advocate—precisely because he had spent the past five years at a large private firm where only the most seasoned lawyers get to argue in court. By the time he rose to the lectern in

Massachusetts, Clark had argued only one prior case in the D.C. Circuit, a minor Medicaid fraud case six years earlier, while in private practice.[7] Everyone was aware that this time he was standing before the D.C. Circuit because of his politics, not because of his demonstrated advocacy skills.

From Clark's perspective, it was entirely appropriate that he should present the oral argument on behalf of the EPA rather than an experienced Justice Department appellate attorney. After all, his political views had played a major role in crafting the legal arguments that the Justice Department had made in its written brief. The EPA career attorneys had wanted the brief to stress what they thought was the agency's strongest argument, one that preserved the agency's discretion to change its mind about how best to address the climate issue. Rather than argue that the plain meaning of the language of the Clean Air Act compelled the conclusion that the EPA lacked authority to regulate greenhouse gases, the EPA career attorneys wanted the brief to argue, relying on the Supreme Court's favorable 1984 ruling in *Chevron v. Natural Resources Defense Council,*[8] that the EPA's interpretation should be upheld because the meaning of the relevant statutory language was ambiguous and the EPA's proposed interpretation, while not required by plain meaning, was at least reasonable. Relying on statutory ambiguity to justify the agency's current interpretation would preserve the EPA's ability at some future date to change to a different reasonable interpretation— including, perhaps, that greenhouse gases were air pollutants.[9] When the meaning of statutory language is otherwise unclear, there can be multiple "reasonable interpretations."

Over the heated objections of career EPA attorneys, the final brief Clark filed with the D.C. Circuit mentioned only in a footnote the possibility of the appellate court deferring to the EPA's interpretation on the ground that the statutory language was unclear.[10] The brief instead relied primarily on the Supreme Court's tobacco ruling in *Brown & Williamson* to support Clark's preferred argument that the only valid interpretation of the Clean Air Act was that it excluded the regulation of greenhouse gases.[11]

Just as the Food and Drug Administration did not have the authority to regulate tobacco in *Brown & Williamson,* Clark argued, Congress could not be deemed to have authorized the EPA to regulate greenhouse gases without strong evidence that Congress had contemplated such sweeping climate regulations when it had passed the Clean Air Act in 1970. The Bush administration wanted to guard against a ruling that upheld the EPA's denial of Mendelson's petition but left the door open for a future EPA to regulate greenhouse gas emissions. Under Clark's direction, the EPA's written brief cited *Brown & Williamson* twenty-two times in thirty-eight pages.[12] Normally the Justice Department does not display such enthusiasm for a prior Supreme Court ruling in which the department lost when arguing in favor of executive branch authority.

Jeff Holmstead, the political appointee at the EPA who had both championed and signed the formal denial of Mendelson's petition, embraced Clark's strategy as his own. After all, the "only reason" Holmstead had recommended that the EPA respond to Mendelson's petition rather than ignore it had been "to lock in the issue" and prevent a future administration from claiming Clean Air Act authority to regulate greenhouse gases.[13]

Clark had played a major role in fashioning the legal arguments of the federal government's brief, but it was still a mistake for him to argue the case himself. Precisely because he had favored making the more challenging argument, the EPA needed to have the argument presented in court by an attorney with more experience in handling the tough questions that would undoubtedly come from the three judges on the bench.

In fact, Clark had overruled the recommendations, including those of other political appointees at the EPA, that a senior career Justice Department attorney should argue the case instead. Their worry was that his participation would "make things too political." It would "conflate policy and politics with legal analysis," undermining the force of their legal arguments. Clark responded that he did not trust the career Justice Department lawyers to do a vigorous job of defending the Bush administration position.[14] Pique, pride, and undue self-confidence got in the way of savvy legal strategy.

Clark's lack of experience was evident during the oral argument. And it was costly. He was repeatedly unable to respond effectively to aggressive questioning from the bench, which began soon after he began his oral argument.

When Clark tried to rely on *Brown & Williamson* at oral argument, he quickly ran into a judicial buzz-saw. As soon as he began his argument, Judge Sentelle cut him off. Sentelle, whom Clark might have assumed would be sympathetic to his legal argument, quickly made it clear that he thought *Brown & Williamson* was irrelevant to the issue then before the court. And when Clark nonetheless referred yet again to *Brown & Williamson,* Sentelle admonished him in no uncertain terms: "Can you forget about *Brown & Williamson?*" he shot back. "Talk about this statute and this case."[15] When Clark inexplicably referred to *Brown & Williamson* again, adding playfully "if I may reuse the forbidden term," Judge Sentelle let Clark have it with both guns blazing:

> I think you might find that if you know that a Court is having trouble relying on a case, you might not want to use that term. Maybe you practice law differently than I did, but both Judge Tatel and I have expressed misgivings as to whether that has a whole lot to do with the case.[16]

Only when prompted by Judge Tatel did Clark abandon his argument that the EPA lacked authority to regulate greenhouse gases and move on to address his "fallback argument"—namely, that the EPA's denial of Mendelson's petition was lawful because the Clean

Air Act did not *require* the EPA "to make a decision one way or the other" concerning whether greenhouse gases endanger public health and welfare.[17] But even then Clark stumbled once more.

In response to a question by Judge Tatel, Clark denied that the EPA was "taking the position that it's free to deny the petition because it has no mandatory duty to decide one way or the other"—even though that was in fact not only exactly what the EPA was arguing, but it was the agency's strongest argument.[18] Perplexed by Clark's response, Judge Sentelle quickly followed up, saying that he thought the EPA was in fact making the very argument Clark had seconds ago told Judge Tatel that the agency was not making. "I thought you were arguing that you had no mandatory duty to exercise that judgment," Sentelle corrected, trying to salvage Clark's case from himself.[19]

When Clark, realizing his blunder, retreated and told Sentelle he was making the argument Sentelle had originally thought the EPA was making, Judge Tatel understandably exclaimed "Now, I'm completely confused!"[20]

None of the oral advocates shone the morning of April 8. The result was a debacle for both sides.

9

A Clarion Dissent

Five weeks after the D.C. Circuit oral argument in *Massachusetts*, the Justice Department's Jeffrey Clark offered a Power Point briefing on climate change litigation to a Bush administration interagency working group on climate change science and technology. *Massachusetts* was one of the five pending cases that Clark described to the working group.[1]

Clark characterized *Massachusetts* as "by far the most important of all the cases" before "the most important of all the courts involved—the D.C. Circuit." He advised those at the meeting that, based on the oral argument, a Bush administration victory seemed likely, with Judge Sentelle being "the deciding vote," while cautioning that "predictions based on oral argument observations are always hazardous." With that caveat, Clark asserted that Judge Randolph "appeared to agree with our defense of EPA's decision on all theories discussed" and that Judge Sentelle, "a conservative, strict constructionist judge," "seemed to agree that the petition to regulate CO_2 from cars and truck was properly denied." By contrast, Clark acknowledged, Judge Tatel, "the panel's most liberal member," "seemed hostile to our defenses on all fronts."[2]

Clark was not wide of mark in his predictions, but none of the parties, Clark included, anticipated the full extent of judicial discordance that would be reflected in the D.C. Circuit's ruling two months later.

A Three-Way Split

The judges met immediately after the morning's oral arguments to deliberate and vote on each of the cases they had just heard. Their shared purpose in those deliberations was to reach a consensus whenever possible. The *Massachusetts* deliberations, however, proved no less of a mess than had the oral arguments. All three judges had very different views on how the court should decide the case.

Judge Sentelle thought the court did not have jurisdiction to hear the case because the petitioners lacked "standing" under Article III of the Constitution, which provides that federal courts can only decide a lawsuit that presents a "case or controversy," which the Supreme Court has in turn ruled means that courts can only decide matters in which the plaintiff bringing the case has "standing." The Court then further described "standing" as meaning that the plaintiff must be able to establish that the defendant has caused the plaintiff an "imminent" and "concrete" "injury in fact" and that the court has the power to "redress" that injury.[3]

Environmental plaintiffs can have a hard time establishing standing because they are often complaining about injuries to the natural environment rather than direct injuries to themselves. In addition, the kinds of environmental injuries they seek to prevent frequently would occur in the future rather than in the immediate present, making them seem less "imminent" and "concrete," more speculative in their causation, and harder for a court to "redress."[4]

Judge Sentelle argued that the petitioners had failed to establish that the EPA's denial of Joe Mendelson's petition would cause them an "imminent, concrete injury" as required by the Supreme Court's prior rulings. Injury from climate change was too attenuated, according to Sentelle, to satisfy Article III—much of it would happen in the distant future as a result of actions occurring over decades across the globe. Because petitioners lacked standing, Sentelle advised his colleagues that the court should dismiss the case without addressing any of the petitioners' arguments that the EPA had acted unlawfully.[5]

Sentelle was well known to be a "hawk" on Article III, but his position was still surprising. No one had seriously questioned the petitioners' standing during the litigation. The EPA had raised the standing issue only in the most cursory fashion in its written brief, but never disputed that the *Massachusetts* petitioners met the injury requirement on which Sentelle now focused.[6] Nor was the lack of attention to the issue in the EPA's brief the result of simple oversight. The EPA attorneys thought the *Massachusetts* petitioners had standing. They were also not normally in favor of making legal arguments that made it harder for environmental groups to bring citizen suits to enforce the nation's environmental laws. The EPA generally supported citizen suits because they provided a useful supplement to the agency's own enforcement actions.[7]

Sentelle had not even raised the standing issue at oral argument, except to follow up briefly on a question from Judge Randolph. Only Randolph had raised the standing issue during the argument, and he had touched on it only quickly. Like the EPA, Randolph had not raised the question of whether the petitioners met Article III's injury requirement. His focus had been elsewhere.[8]

Randolph agreed with Sentelle that the *Massachusetts* petitioners should lose the case, but on an entirely different ground. He did not want to question their Article III standing, and he also did not want to address the question of whether greenhouse gases were air pollutants within the meaning of the Clean Air Act. He favored bypassing both of those questions and addressing instead the EPA's backup argument that the agency had discretion to deny Mendelson's petition on the ground that it did not yet want to decide on the endangerment issue. Because Randolph thought the EPA clearly possessed such discretion, he argued that there was no good reason for the court to address either of the two harder issues.[9]

Tatel disagreed with both his colleagues. He thought the *Massachusetts* petitioners clearly had Article III standing, that greenhouse gases no less clearly were "air pollutants" within the meaning of the Clean Air Act, and that the EPA had acted unlawfully in declining to consider whether emissions of such gases from new motor vehicles endangered public health and welfare.

The disparate views of the three judges meant there would be no official "opinion of the court." An opinion of the court requires that a minimum of a majority of two agree on a common rationale. Under the exceedingly rare circumstance of there being no majority opinion, the court would be relegated to issuing a "judgment" upholding the EPA's denial of Joe's petition—followed by two separate opinions by Judges Sentelle and Randolph expressing their different reasons for coming to that conclusion, neither one of which would establish legally binding precedent. Judge Tatel would have the right, if he wanted, to file a dissenting opinion, explaining why both of his colleagues were wrong.

"Make It Long"

The result of the court's private deliberations was not good news for the *Massachusetts* petitioners, although none of the parties would know that until months later when the court's opinion was published. The vote was two to one, in favor of the EPA. All that needed to be done now was for the majority and dissenting judges to draft their opinions, explaining the result and their disparate rationales.

As Tatel returned to his chambers following those private deliberations, he decided he would file a dissenting opinion.

For some judges, the filing of a dissenting opinion would have been an automatic decision. After all, they had been outvoted and the court's ruling would not reflect their views of the case. But not for Tatel, who rarely dissented even when, as had just happened, he disagreed with the other two judges in a case. Out of respect for his colleagues, Tatel publicly dissented only if he concluded it was truly important to do so.[10] At the time of *Massachusetts*, there were dissenting opinions filed in fewer than 3 percent of all D.C. Circuit cases.[11] Tatel himself had published only one dissenting opinion in more than two years.[12]

The case had completely fascinated Tatel. The son of a highly regarded research physicist, himself originally a physics and math major at the University of Michigan—before the social unrest of the 1960s drove him to take classes that he perceived as more "politically relevant"—he loved cases with science in them.[13] Back at his chambers following the morning deliberations, Tatel called his "clerk for the day" into his private office to discuss next steps. His practice was to assign one of his four law clerks—all recent graduates from top law schools—to all the cases argued on any one day. The judge outlined to his clerk what he wanted the first draft of his dissenting opinion in *Massachusetts* to say: what issues his opinion should cover, what conclusions it would reach, and what reasons it would stress in support of those conclusions. And he told the clerk "to make it long." The clerk should be sure to cover every issue.[14]

The judge was also very firm about both how the opinion should begin and how it should end. He viewed the beginning and the end of the opinion as its important "framing pieces." The beginning should be a clear and cogent explanation of the legal issues raised in the case and the views of his two colleagues on the bench with whom he disagreed. The end should strip the case of its political overtones and make it clear that its resolution turned on nothing more than a straightforward application of well-settled principles of administrative law.[15]

Tatel also explained that the intended audience for his dissenting opinion extended to the Supreme Court Justices. If the *Massachusetts* petitioners decided to seek Supreme Court review, Tatel wanted any Justice on the fence about whether to hear the case to be persuaded by his dissent both that Supreme Court review was appropriate and that the Justices should overturn the EPA's denial of Mendelson's petition. The odds of the Supreme Court deciding to grant review in the case were small, but the stakes, Tatel thought, were as large as in any case he had faced in his decade on the bench. Tatel later put it to his judicial colleagues who did not join his dissent, "If global warming is not a matter of exceptional importance, then those words have no meaning."[16]

Tatel's clerk was responsible for preparing the first draft of the opinion. But like every Tatel clerk, the clerk assigned to draft the *Massachusetts* opinion knew it would not be the final draft. Tatel threw himself into the opinion writing process. As one former clerk described it, it was only after the clerk provided the judge with the first draft that "things started to get fun."[17]

Tatel routinely worked through twenty to fifty draft opinions in a case before agreeing to a final version. The *Massachusetts* case was a fifty-draft opinion. Tatel immersed himself in every aspect of the case, including the science of global warming, which he found fascinating. He read and reread a National Academy of Sciences report on the science of global warming and climate change and worked around the clock for several months on draft after draft.[18] His clerk spent many long days and nights at the office, and stayed overnight at least once. The last thirty drafts all involved precise word choices and an effort to add more "color" and "oomph" to the opinion.[19] There was a good reason Tatel had a reputation throughout both the D.C. Circuit and down the street at the Supreme Court as a "judge's judge"—even those who might disagree with his conclusions on a particular case admired his analytical mastery.

Tatel hated to be reversed by the Supreme Court. All the legal analysis in his opinions had to be tightly and exactly reasoned and supported by authority. Nor did he want his published opinions to have any typographical or grammatical errors. Opinions were proofed and reproofed to guard against any possible mistake. No Tatel clerk wanted to disappoint the judge by letting one slip by.[20]

Part of the process entailed Tatel and his clerk reading the draft opinions out loud. His clerks marveled at his memory. He seemed to know every word of every paragraph. If one read the first few words of any paragraph, he could recite verbatim the rest of the paragraph. As hard as the clerks worked, they knew that Tatel was working even harder, not because of his disability but because he loved every minute of the process.[21]

The clerks often forgot that Tatel was blind. He had lost his sight soon after law school when he contracted the eye disease retinitis pigmentosa. He had to make accommodations in both his personal and his professional life to address his lack of vision. When he ran marathons, he would be linked to a running partner by a heavy-duty shoestring attached to his upper left arm. When he skied downhill, a skiing partner, typically one of his children, would ski close behind him and communicate by radio transmitter about upcoming turns and obstacles.[22]

For *Massachusetts,* as in all his cases, Tatel employed a reader, a recent college graduate, to read the briefs and other documents out loud. The reader's job was to read very quickly so as not to waste the judge's time. When written material was in Braille, Tatel would use his Freedom Scientific Braille 'n Speak machine, which would convert written Braille into spoken speech. Tatel's children referred to it as "Dad's Fisher-Price toy."[23] Tatel would set the device to triple speed so he could consume material more quickly. The speech was so fast that his clerks had trouble processing it.[24] Tatel regularly complained about the voice, which he described as sounding like a "Swedish aunt."[25]

Many attorneys who argued before Judge Tatel did not know that he was blind. And for good reason: it never affected his ability to do his job. Apart from a small earpiece in his ear, which allowed him to review his notes and hear excerpts from relevant statutes or regulations, there was absolutely nothing to suggest the judge's physical disability. And even for the attorneys who knew, Tatel's knowledge of the precise wording of language at issue in the case made it easy to forget. Tatel would regularly correct advocates who misquoted relevant language, advising them that was not the language that he "saw" when he read it.

On July 15, 2005, three months after the oral argument, the D.C. Circuit issued its opinion in *Massachusetts v. EPA*. Consistent with the private deliberations held immediately after oral arguments, the vote was two to one in favor of the EPA and there were three separate opinions. Sentelle and Randolph supported dismissal of Mendelson's petition, though for different reasons. Alone, Judge Tatel dissented.[26]

Judge Randolph's opinion, six pages long, assumed the *Massachusetts* petitioners had satisfied Article III standing and bypassed the question of whether greenhouse gases were air pollutants. It reasoned that the EPA's denial of Mendelson's petition amounted to a valid exercise of the broad discretion the Clean Air Act had conferred on the agency to exercise its "judgment" in considering such petitions.[27]

Sentelle's opinion addressed only the Article III standing issue. It took the judge little more than two pages to explain why the petitioners had failed to demonstrate the requisite injury to themselves to bring the lawsuit to federal court.[28]

Judge Tatel's dissent was more than twenty pages long, and it detailed all the legal issues raised in the litigation. His opinion included a lengthy background discussion of climate change science with repeated references to National Academy of Sciences reports. The dissent made it clear that Tatel thought that both the question of standing and whether greenhouse gases were air pollutants were relatively straightforward legal issues for which the petitioners clearly had the stronger position. Addressing "what EPA's counsel terms 'the fallback argument,'" Tatel acknowledged that an agency normally has significant discretion to deny a petition that, like Mendelson's, asked an agency to adopt new regulations before the agency thinks it is the best time to do so. And a court should be prepared to defer to the agency's judgment. But here, Tatel concluded, it was "difficult even to grasp the basis for the EPA's action." The agency's published decision suggested one possible basis and its written brief filed with the court suggested another basis. Because Tatel found both unpersuasive, he concluded that he did not need to determine which of the two possible bases was its actual justification: "It is obvious that none of EPA's proffered policy reasons justifies its refusal to find that [greenhouse gas] emissions" endanger public health and welfare.[29] None of the EPA's reasons, Tatel stressed, even addressed this central issue. This, to his mind, was clearly arbitrary and capricious and therefore should be struck down as unlawful.

"Leave It Alone"

On the morning of July 15, the D.C. Circuit clerk's office called the counsel for both the *Massachusetts* petitioners and the EPA to notify them that the court had issued its ruling. Not surprisingly, the attorneys at the Justice Department and the EPA were pleased by the court's bottom-line upholding the EPA's action. The counsel for the petitioners were disappointed, but not surprised. They had always known it would be an uphill battle to persuade a court to override a federal agency's determination to decide *not* to decide something. Once they had learned that Sentelle and Randolph were on the three-judge panel deciding the case, their concern about whether they could win the case turned into dread of what the effect of a major loss could be.

When they finally had a chance to read the opinions, many of the attorneys for the *Massachusetts* petitioners were privately more relieved than disappointed. Although they had lost, they had not lost in a way that had established precedent that would limit their ability to bring future climate litigation. In fact, the appellate court had not established any precedent at all, because Randolph and Sentelle had not agreed on why the EPA should win. Given the possibility that those two judges might have instead joined an opinion that denied environmentalists Article III standing in all climate cases or that held greenhouse gases were not air pollutants, the petitioners had reason to breathe a collective sigh of relief. They had dodged a judicial bullet.

That is why, when the many co-counsel had a conference call five days later, on July 20, to discuss next steps, there was almost universal agreement that they should not press the case further. There was no good reason to seek rehearing by the panel or a new hearing by the full D.C. Circuit, or to seek Supreme Court review. Best just to "leave it alone," and move on to the next, and potentially far better, opportunity to raise the climate issue.[30] The environmental organizations were unanimous in that view. Even powerful states like New York thought it was best to fold up this particular tent and wait for the next case—preferably one in which New York might be the lead state.

One Carbon Dioxide Warrior, however, disagreed. He was, perhaps unsurprisingly, the lead lawyer for the Commonwealth of Massachusetts. Jim Milkey was not ready to give up the fight.

10

Hail Mary Pass

The future of the environmental movement is on your head."[1] Those were the chilling words that Frances Beinecke, the highly regarded president of the Natural Resources Defense Council, spoke to Jim Milkey in late August 2005, in an effort to persuade him to abandon his plan to seek reversal of the D.C. Circuit's ruling in favor of the EPA. Beinecke called Milkey after first taking the extraordinary step of sending a formal letter to Milkey's boss, Massachusetts attorney general Tom Reilly. "I am writing with great urgency to urge you not to take this case any further," she told Reilly. Seeking further review, Beinecke explained, in both her letter and her phone call to Milkey, risked "disastrous" consequences that would set back all their efforts to address global warming. Beinecke closed by urging Massachusetts to let this case go: "We gave it a good shot in this case. We didn't win, but we didn't lose either. And now this is no longer the best battleground."[2]

Beinecke's intervention was the final confrontation between Milkey and the Carbon Dioxide Warriors led by Doniger, who had splintered into factions after the court ruling. No one disputed that the odds of overturning the D.C. Circuit ruling were very small. Or that pursuing further judicial review threatened a worse outcome. Even if the full D.C. Circuit agreed to hear the case, there was a chance that the petitioners would lose in a way that would cause far more harm to future climate litigation than the splintered, nonbinding opinions of Judges Randolph and Sentelle. Beinecke's letter, which Doniger undoubtedly helped craft, put the odds of a favorable result as "next to zero."[3]

But Milkey was determined to press on. "If we are scared to pursue litigation on the grounds we may lose," he challenged the others, "haven't we already lost?" He also expressed great concern "about the public message that our dropping the case could bring. . . . [T]he headline is going to be that we gave up." Milkey also reminded the others that Massachusetts was no longer "collectively locked into having to reach consensus" on whether to seek further review. Massachusetts was free on its own, regardless of what the others might conclude, to decide to go it alone.[4]

None of the environmental groups agreed, and the reaction of the other states was at best reluctantly tepid so as not to disrespect Massachusetts. Doniger pushed back the hardest. He stressed that further review would risk an adverse court ruling that could seriously damage their prospects of bringing future environmental cases of all stripes—not just those related to climate change—to federal courts. They could lose on Article III or on the meaning of the term "endangerment" in a way that could drastically cut back on the right of private citizens to bring lawsuits forward claiming harm from pollution.[5] Doniger was incensed by Milkey's suggestion that, regardless of what the other petitioners decided, Massachusetts would file on its own, along with anyone else who'd like to sign on. That is when he decided to bring in the big guns and to have his own boss write such a strongly worded letter to Milkey's boss.

Doniger's position on behalf of NRDC was heavy-handed but understandable. NRDC had been a leading environmental public interest litigator for decades—its motto was "The Earth's Best Defense"—and an early champion of the climate issue. The litigation stakes in *Massachusetts* were huge, which is why Doniger had fought hard to maintain control over the litigation, including by insisting that the environmentalists and the states file a single joint brief. Now he risked losing control over the litigation because of Milkey—who was threatening to take action on his own that Doniger believed could have disastrous consequences for NRDC's important work.

Doniger succeeded in having the environmental groups unified in their opposition to Milkey. Mendelson did not dissent, even though he had previously been willing to buck the national groups when they had tried to prevent him from pressing his case at the outset. He, too, agreed it was time to fold this particular tent.

But Attorney General Reilly—born, raised, and educated entirely in Massachusetts—was not receptive to NRDC's pleas. After eight years in office, he was not prone to second-guess the judgment of his senior career lawyers, especially at the request of a national public interest group with which he had never himself worked closely and that lacked any distinct Massachusetts presence. He refused to overrule Milkey's recommendation, and Massachusetts filed a petition for rehearing before the full D.C. Circuit, which a few states and none of the environmental organizations joined. The full D.C. Circuit denied the petition to rehear the case by a vote of four to three, with two of the nine active judges not participating. As expected, Judge Tatel dissented on the ground that global warming clearly presented a matter of exceptional importance warranting the full court's review.[6] Two other D.C. Circuit judges who had not served on the three-judge panel joined Tatel in favor of rehearing, but that was still one vote short of the majority necessary to reopen the case. If there was to be a next stop, it would have to be the Supreme Court.

Cert

To persuade the Supreme Court to hear any case is a Herculean undertaking. Unlike lower federal courts, which must hear all appeals, the Justices have virtually unbridled discretion to decide whether to hear a case. They receive 6,000 to 8,000 requests each year to review lower court rulings. And they agree to hear and decide only 65 to 75 of these cases.[7]

The likelihood of a hearing is vanishingly small. It's not that the Justices believe the lower courts make few mistakes. They are well aware that many of those thousands of cases—at least in the hundreds—were decided incorrectly. But they all agree that correcting lower court errors is not their job. They have a far more exalted understanding of why the Framers of the Constitution wanted to establish a "Supreme Court" and of their related responsibility as Justices on that Court. Their constitutional assignment is to reserve their time and attention for deciding the most pressing legal issues facing the nation. As one Justice pointed out, the quickest way for Congress to make the Supreme Court ineffective would be to "bury it" with cases by depriving the Justices of their current ability to decide to hear only a small fraction of the cases for which review is sought.[8]

For the Justices, the touchstone for when the Supreme Court should grant review is not whether the lower court ruling was correct but instead whether the legal issue decided by the lower court is of such overriding importance that it should be answered by the Supreme Court. Sometimes that will be when the lower court ruled incorrectly, but not always. Supreme Court review can sometimes be called for if the Justices believe the lower court ruling was correct. Or if there are multiple lower court rulings in conflict with one another and the country needs one uniform answer to govern everywhere.

Immediately after the D.C. Circuit denied the rehearing, Milkey set his sights on the Supreme Court. He weighed the pros and cons of seeking High Court review and recommended in favor of filing.

He reasoned that there was no "appreciable downside to the larger public debate of potentially fighting the good fight but losing."[9] He had surveyed the other governmental parties and environmental organizations for their views on the wisdom of seeking such review, but there was never much question that Massachusetts would file a "petition for a writ of certiorari"—or a "cert petition"—the formal document asking the Court to review a lower court's decision.

This time the environmental organizations did not repeat the scorched-earth tactics they had unleashed to try to stop Milkey and Massachusetts from seeking further review before the D.C. Circuit. That had left deep battle scars among the Carbon Dioxide Warriors—the harmony they had enjoyed at the outset of the litigation had now largely disintegrated. That battle had left no doubt about Massachusetts's resolve or its willingness to go it alone if necessary. Doniger quickly decided it would be far better to help shape Milkey's cert petition than to fight its filing and lose all control over its contents.[10]

Doniger and others who worried about Supreme Court review knew the absurdly long odds facing Milkey's cert petition. The chance of the petition being among the roughly 1 percent of the cases the Supreme Court would choose to hear was close to zero.

Milkey had little Supreme Court litigation experience, but he knew that the case was the longest of long shots. His problem? The D.C. Circuit had ruled against the *Massachusetts* petitioners, but it had done so without making any authoritative statement of law. The only thing about which two judges had agreed was that the EPA should win. There was no overlap in their reasoning, and therefore no legal principle or precedent had been established for future cases. Milkey was hard pressed to argue that the Court's review was required to address an important issue of federal law decided by the lower court. Any such claim bordered on the disingenuous.

The D.C. Circuit had ruled in favor of the EPA without answering any of the important questions the case had raised. Only Judge Sentelle had concluded that the *Massachusetts* petitioners lacked standing to bring the case. Only Judge Randolph had concluded that the EPA had the discretion to deny Mendelson's original petition. And neither of those two judges had bothered to address the central issue in the case: whether greenhouse gases were air pollutants. In these circumstances, as any expert Supreme Court advocate would have advised Milkey and his co-counsel, denying review would be a no-brainer for the Justices. Supreme Court Justices do not believe it is their job to rule on legal issues that have not yet been decided by the lower courts.

Against these odds, Milkey nonetheless gamely drafted a cert petition for Supreme Court review, which he circulated in late January 2006 to all the other counsel for the petitioners who had joined together in the D.C. Circuit litigation.[11] The good news was that the draft resulted in a quick consensus. But the consensus was that Milkey's draft wasn't very good.

Milkey's draft had begun by blasting Judges Sentelle and Randolph for shirking their judicial responsibilities. It accused Sentelle of bias and claimed that his resolution of the case had been so "unprincipled" as to call into question the justification for an independent judiciary as envisioned by the Framers of the U.S. Constitution. Both Sentelle and Randolph, the draft argued, had "essentially repudiated judicial review" and "mock[ed] that purpose."[12] The gist of the draft cert petition's complaint was that there was something untoward about the EPA's winning when the agency had failed to receive a majority of votes on any of the three legal issues before the D.C. Circuit.

Milkey's strategy was to take an obvious weakness in his case—the lack of a ruling on any important question of federal law—and try to convert it into a strength. This approach had one merit: it didn't pretend the problem wasn't there. But attacking federal judges in front of other federal judges is, at best, a foolish gambit. It is more likely a quick way to crater your case even if it otherwise has merit. As one commenter put it to Milkey, the draft's "attack on the D.C.

Circuit is a . . . disaster. If the Justices think they can get to the issues you present only by dumping all over Randolph and Sentelle in the way you suggest, this petition is done."[13]

To his credit, Milkey backed down. Not only had his draft "failed to identify a compelling legal theory," but he conceded that "arguments aimed at the behavior and underlying discretion of judges are probably doomed to failure (especially when the judges we attack are literally friends of the judges we are trying to persuade)."[14] He then made a game-changing decision. This time Milkey did not double down and attempt to redraft the petition himself. Instead, he brought in reinforcements. He turned to Lisa Heinzerling—a former colleague who had left the Massachusetts Attorney General's Office to become a law professor at Georgetown University—and asked her for help.

The original "five guys" now had a sixth—a gal—and she transformed the case.

A Pistol from Minnesota

Lisa Heinzerling was in her early forties, but would easily pass for a decade younger, and was widely celebrated as one of the nation's leading environmental law scholars. She was also by all accounts a spectacular, highly engaging classroom teacher. She possessed impeccable academic and professional credentials, as well as a biting wit and an infectious smile. An honors graduate of Princeton University, Heinzerling had been the first woman ever to serve as editor in chief of the *University of Chicago Law Review*. Following law school, she had clerked for Justice William J. Brennan Jr. of the United States Supreme Court.

A clerkship with a Supreme Court Justice is, by leaps and bounds, the most prestigious position there is for a recent law school graduate. About forty thousand students graduated from law school in 1987, Heinzerling among them, and only thirty-six of them went on to

Supreme Court clerkships. Those who managed to grab the brass ring were the top students at the nation's top law schools, many of whom went on to become leading lawyers, government counsel, or law professors at prestigious universities. Five members of the current Supreme Court had clerked for Supreme Court Justices.

Like many of her generation, Heinzerling could recall the impact of the nation's first Earth Day, in April 1970, when she was only nine years old. Growing up in rural Minnesota, she had always loved the outdoors. In her family, "it was kind of a sin to go a whole day without going outside. . . . When you grow up like that," she wrote, "you do not learn to think of yourself as a thing separate from nature."[15]

The schoolroom discussions back then were not, of course, about climate change. The environmental issues that gripped people in rural Minnesota in the early 1970s were pesticide contamination and gasoline spills from motorboats harming the beautiful lake that the Heinzerling family enjoyed, and upon which the wildlife they cherished depended.[16]

At that time, terms like "climate change" and "global warming" were decades away from being known to the general public. Atmospheric concentrations of greenhouse gases were at a comparatively benign 327 parts per million, and only a small handful of relatively unknown scientists were studying in earnest the impact of increasing carbon dioxide emissions on the climate.[17] Whatever alarm bells might have otherwise rung in the federal government, as atmospheric greenhouse gas concentrations began to rise precipitously, were stilled by the election of Ronald Reagan in 1980.[18]

As a young lawyer, Heinzerling had a clear commitment to social justice. Justice Brennan nicknamed her "Pistol" because of her fiery personality and take-no-prisoners writing style, and her fellow Supreme Court clerks voted her the "least likely to make a million dollars" in recognition of her commitment to public interest work. She wasn't much interested in pursuing the lucrative private law firm salaries and signing bonuses readily available to Supreme Court clerks.[19]

Following her clerkship with Justice Brennan, Heinzerling was one of the few clerks who did not go to a private law firm. She instead accepted a fellowship, which paid about 60 percent less than a law firm salary, to work for a public interest organization in Chicago that focused on energy and consumer protection issues, before accepting an environmental law job working for Milkey in the Massachusetts Attorney General's Office. She worked with Milkey for three years before joining the Georgetown law faculty in 1993.

When Milkey turned to Heinzerling in early February 2006, it was not the first time Massachusetts had sought to enlist her help in a Supreme Court case since she had left the Attorney General's Office. Seven years earlier, the state had lost another Clean Air Act case in the D.C. Circuit and had decided to seek Supreme Court review. Because Heinzerling had clerked for Justice Brennan and had insider expertise that the Massachusetts Attorney General's Office otherwise lacked, she was asked to serve as the Commonwealth's "Counsel of Record" in petitioning for Supreme Court review.

But that case, *Whitman v. American Trucking Associations, Inc.*,[20] had been a cakewalk compared to *Massachusetts v. EPA*. In that case, Massachusetts had been aligned with the Clinton administration, and the EPA had filed its own petition seeking the Court's review. The Supreme Court almost always grants review when a federal agency advises the Court that a case is sufficiently important to warrant the Justices' time.[21] Once a federal agency files a cert petition, whether states file their own petitions is largely beside the point. And, in fact, that was what happened in *Whitman*. The Justices granted only the EPA's petition in *Whitman v. American Trucking*, postponing any consideration of Massachusetts's petition until after it had decided the EPA's case.[22]

Milkey was now asking Heinzerling to help him do something no lawyer had ever successfully done before. By 2006 the EPA had been

in existence for more than thirty-five years. During those four decades, environmentalists—sometimes supported by states, and sometimes on their own—had frequently sought Supreme Court review
of their lower court losses to the EPA. How many times had the
Court granted review in response to their requests? Zero.

Looking beyond the EPA, in the Court's more than two-hundred-
year history the Justices had only *once* before ever granted review
at the request of environmentalists over the federal government's
opposition. The single, isolated case was in 1971, when the Justices
agreed to hear a case brought by the Sierra Club against the U.S.
Department of the Interior.[23] That victory proved short-lived. Although the Sierra Club made history by persuading the Court to
hear the case, the Court promptly ruled against the Sierra Club by
a vote of seven to two.[24]

Heinzerling had less than four weeks to pull off what seemed like
an impossible feat. Although a party has ninety days to seek Supreme
Court review after losing in a lower court, Massachusetts had already spent more than sixty of those ninety days in conference calls
and fielding emails, reviewing Milkey's now-discarded draft. Heinzerling had less than a month to come up to speed on a case that others
had been litigating for years, do the necessary additional research,
draft a new petition, reach the kind of consensus on its contents that
had so far proven elusive with so many lawyers involved, and file a
cert petition with the Court that had been thoroughly checked and
double-checked for any possible mistakes. It was not for the faint of
heart. But Heinzerling was not one to be put off by a challenge.

She had the opening advantage of being both trusted and highly
regarded by Milkey, as well as by environmentalists at NRDC,
Earthjustice, and the Sierra Club, who knew her to be an unabashed
environmentalist. Several of her former law students had gone on
to work for the leading environmental groups involved in the litigation, and Heinzerling's own scholarly writings left no doubt that she
shared their passion for environmental protection. A book she had

written with an environmental economist and just published the previous year, *Priceless: On Knowing the Price of Everything and the
Value of Nothing,* was widely praised by environmentalists. The
book pulled no punches when it came to expressing outrage at those
who would try to put a price tag on the loss of life or serious illness
caused by pollution to justify cutting back on environmental protection requirements.[25]

Heinzerling was also good friends with the Sierra Club's Bookbinder. They had known each other for decades, beginning with
their overlapping years at both Princeton and the University of Chicago. Heinzerling had even introduced Bookbinder to Milkey and
the Massachusetts Attorney's General Office when he had wanted
to leave his New York City law firm.

On February 16, less than two weeks after bringing Heinzerling
on board, Milkey circulated her revised cert petition to all the parties who had joined Massachusetts in its litigation in the D.C. Circuit. He gave them only eight days to decide whether to sign on. His
email included a blunt warning: Heinzerling had "profoundly
alter[ed] the thrust of the petition," he wrote; and "some of the
changes can be fairly shocking to anyone who has been deeply immersed in the case." Milkey then added that he fully endorsed the
new draft because he was persuaded that Heinzerling's changes "will
make the Supreme Court much more interested in accepting the case
(understanding, of course, that this is always a crapshoot)."[26]

So, what, exactly, had she done?

First, she came up with a new way to pitch the case to the Supreme Court. She focused the petition on the legal issue that she felt
would be the most interesting to the Justices, even if it was not the
one that was most important to Massachusetts, the environmentalists, or any of the other petitioners on their side. Having herself reviewed many cert petitions as a law clerk, Heinzerling had a keen
sense for the kinds of legal issues most likely to pique the Justices'
interest.

The question of whether greenhouse gases were air pollutants—the issue the petitioners cared the most about—nearly vanished behind the kind of arcane, technical question of administrative law that is the daily diet of Justices: what discretion does an administrative agency have to decide to postpone making a decision that Congress has authorized the agency to make? This question may sound boring or pedantic, but it was exactly the kind of potentially sweeping question of administrative law—governing the relations between executive branch agencies and the courts—that makes a Justice's heart beat faster.

The greenhouse gas issue, by contrast, concerned the meaning of only one word in one statute applied in one discrete context that the Justices knew little about. No Justice was a dogged environmentalist looking for opportunities to use the Court's docket to promote stronger environmental protection the way some Justices, like Thurgood Marshall and William Brennan, had clearly embraced their role on the Court to safeguard the civil rights of racial minorities, or Ruth Bader Ginsburg to fight gender discrimination. Even more, neither D.C. Circuit Judge Sentelle nor Randolph had even addressed the greenhouse gas issue in their two separate opinions, and Justices never like to be the first to address a legal issue in a case.

So, knowing all this, Heinzerling cleverly reversed the order of the legal issues in her revised cert petition and put the issue of administrative law first and the greenhouse gas issue second, even though, as matter of strict logic, the Court should get to the issue of whether the EPA had the discretion not to regulate a pollutant over which it had authority only *after* considering whether the agency had the authority to regulate greenhouse gases at all.[27] Living up to her nickname, Heinzerling put a bull's-eye on Randolph's opinion, which she dubbed the "lead opinion"—a title entirely made up by her in an effort to distract everyone from the problem that there was no lower court majority opinion. She fired at Randolph's rationale, arguing that, if left to stand, it "would be to sanction an enormous shift of power to administrative agencies, effectively letting them dismantle statutory regimes they simply did not like."[28]

Heinzerling's cert petition deliberately submerged the environmental dimension of the case. It raised the separate question whether greenhouse gases were air pollutants only briefly and secondarily—long after the extended discussion of the administrative law issue upon which, the cert petition intimated (with somewhat exaggerated prose, such as its suggestion that agencies could "dismantle statutory regimes they simply did not like") the fate of the nation depended. As Milkey warned his co-counsel in his email attaching the new draft and soliciting their agreement to sign on, the new version's "focus is on the importance of the questions presented as a matter of the Supreme Court's overarching teachings." He knew this would be hard to swallow for the environmental groups. "In fact," he wrote, "for most of the petition, you would hardly know that the case was about global warming." But, Milkey added, this was "obviously done" because to win this case they would "need to win the votes of conservatives justices who are not likely to be inclined in our favor on the underlying controversy."[29]

Heinzerling's draft made clear that she appreciated she could signal to the Court the importance of the case by taking advantage of the high regard in which the Justices held Judge Tatel, who had dissented in the D.C. Circuit. About one-third of the current Supreme Court clerks had clerked the prior year on the D.C. Circuit when *Massachusetts* had been decided, including two for Judge Tatel, when he had written his dissenting opinion. Those clerks would be familiar with the case and the rarity of a Tatel dissent.

Heinzerling's revised cert petition fully embraced Tatel's dissenting opinions (both his initial opinion and his subsequent opinion dissenting from rehearing). Emulating Tatel, her draft acknowledged that the EPA normally possessed great discretion to decline to grant a petition to regulate. "But, as Judge Tatel observed," the draft specified, "the statute provides the Administrator 'no discretion either to base that judgment on reasons unrelated to [the statutory] standard or to withhold judgment for such reasons.'" Quoting directly from Judge Tatel, the draft petition argued that the EPA did not possess the discretionary authority "to withhold regulation because it thinks such regulation bad policy. . . . Congress did not give EPA this broader authority, and the agency may not usurp it." The cert petition was the advocate's equivalent of a group hug of Judge Tatel.[30]

Unlike many, Heinzerling wrote with flair and style. Too many lawyers mistakenly believe that legal prose must be mechanical, dispassionate, and rigidly formal in tone. Heinzerling understood that she needed to grab the Justices' attention, to make it clear that the *Massachusetts* case was special and not just another one of the thousands of well-intentioned cases destined for the Court's routine denial of review. The best Supreme Court advocates appreciate that the Justices and their law clerks, otherwise saddled with reading hundreds of pages of potentially dull prose, do not mind occasionally being entertained. Justice Scalia's book on "The Art of Persuading Judges" admonishes lawyers to "make it interesting" by being "more livid, more lively, and hence more memorable" in their writing.[31]

Heinzerling's legal argument confidently began with a literary analogy far removed from the seriousness of the separation of powers and the portent of catastrophic climate change: "Like Melville's scrivener, Bartleby, EPA may as well have explained its resistance to fulfilling its statutory obligations under the Clean Air Act with the simple reply: 'I would prefer not to.' And, like Bartleby's flummoxed boss, the court below let this answer suffice."[32]

The petition glistened with personality and clever turns of phrase. According to Heinzerling, the D.C. Circuit's decision was "seriously out of step with this Court's precedents counseling judicial modesty in the face of pellucid statutory language." The meaning of the relevant Clean Air Act language was not merely clear, Heinzerling teased, it "is crystalline" and the EPA had "belittled the very idea of textual analysis." An ordinary cert petition would have asked the

Court to grant review to correct a "legal error." Heinzerling's petition stated that the petitioners were seeking "this Court's review to slip the case back into its proper legal joint." Her petition turned what had the potential to be mind-numbingly dull analysis of highly complex statutory language—weighed down by jargonistic references to "emissions standards," "requisite technology," "useful life," "complete systems" and "incorporate devices"—into a clear, accessible, and simple presentation. It was not by accident that the words "simple," "simply," and "simplicity" appeared eleven times in the petition. The petition was a tour de force in demonstrating how sophisticated legal analysis can be combined with engaging, straightforward, and readable legal prose.[33]

The conflict that had plagued the cert petition's drafting vanished almost as quickly as Heinzerling's new draft was circulated to the many co-counsel for their review and approval. There were incidental debates and discussions about this or that, but with the filing deadline fast approaching, the consensus that had eluded *Massachusetts* petitioners in the D.C. Circuit briefing, and then broken down completely in the bitter disagreement over whether to seek rehearing *en banc,* was now quickly reached. The petition for a writ of certiorari was successfully filed with the Supreme Court on the March 2 deadline, with all the original petitioners now on board.

Milkey and Doniger independently described as "brilliant" Heinzerling's tactic of "inverting the questions presented" by putting the administrative law issue up front.[34] After months of destructive infighting, our two protagonists had finally found a new common ground, thanks to their new Carbon Dioxide Warrior. Heinzerling had successfully pitched the case as well as could be done. The odds that the Supreme Court would grant review still remained vanishingly small.

11

"Holy #@$#!"

The Justice Department's Office of the Solicitor General of the United States, which represents all federal agencies in cases before the Supreme Court, received the *Massachusetts* cert petition by mail on March 7, 2005. Under the Court's rules, the due date for the federal government's filing a brief in opposition to any petition is nominally thirty days, but the federal government routinely seeks and receives one, two, or even three extensions due to the press of business the solicitor general has before the High Court. In *Massachusetts,* the Court granted the solicitor general three extensions, until May 15, which is the day the federal government's brief was finally filed.

The solicitor general of the United States at the time was Paul Clement, who at thirty-eight years old was already regarded as one of the nation's most able Supreme Court advocates. A graduate of Georgetown University and Harvard Law School, Clement enjoyed the same academic pedigree as Supreme Court Justice Antonin Scalia, for whom he had clerked after graduating with high honors from law school. Thanks to his outstanding skills as a lawyer, and his personal charm and professional collegiality, Clement succeeded in being widely admired by both political appointees and career attorneys within the Justice Department and across the federal government. The Solicitor General's Office under Clement's stewardship was known for the excellence of its work. The *Massachusetts* petition, upon arrival, should have seemed like a routine cert denial. But the case kept on defying expectations.

The Office of the Solicitor General

There are no more skilled Supreme Court advocates than those who work in the Office of the Solicitor General. A position first established by Congress in 1870, the solicitor general must be "learned in the law."[1] The first solicitors general championed the enforcement of civil rights law against those who resisted the full emancipation of slaves following the end of the Civil War.[2] Today the solicitor general is best known for having exclusive authority to represent the United States in the Supreme Court.

Any federal agency that has lost a case in a federal court of appeals can seek Supreme Court review only if the solicitor general agrees. Just as the Justices can decide not to hear a case, the solicitor general can refuse any agency's request to seek Supreme Court review. The solicitor general can even confess error in the Supreme Court and, over the objection of the government agency in question, tell the Court that a decision that the federal government won in the lower courts was wrongly decided. Unlike other clients represented by legal counsel, federal agencies can't just go out and hire another lawyer to represent them if they are dissatisfied with the solicitor general. They are stuck with the solicitor general's decision, unless they can either persuade the solicitor general to permit them to hire their own lawyer or persuade the president of the United States to order the solicitor general to decide differently, something that is not easily done.

The solicitor general is supported by a small staff of about twenty attorneys, four of whom are deputy solicitors general and sixteen assistants to the solicitor general. Their small numbers belie the depth of their Supreme Court expertise. All are top graduates from the nation's law schools who within years of graduating have already established themselves as outstanding appellate advocates, many clerked for Supreme Court Justices, and they have intimate knowledge of the internal workings of the Court and the predilections of the individual Justices.

At the time of *Massachusetts*, the two most recent appointments to the Supreme Court—Chief Justice John Roberts and Justice Samuel Alito—had previously worked in the Solicitor General's Office. Roberts had served as principal deputy solicitor general and Alito as an assistant to the solicitor general. Several subsequent Court appointees had also worked in the office, including Justice Elena Kagan, who was solicitor general, and Justice Brett Kavanaugh, who served in 1992 as a Bristow Fellow in the office when he was twenty-seven. Before Kagan, four additional solicitors general went on to serve as Supreme Court Justices, one of whom was Thurgood Marshall. Another, Chief Justice William Howard Taft, also served as president of the United States.

Only two of the attorneys in the office—the solicitor general and one deputy—hold "political positions," meaning that those appointments can change with every presidential election. All the others are nonpolitical career positions, and many of the attorneys stay for years, across presidential administrations, and sometimes even for decades. The core strength of the office derives from the longevity of the career attorneys and the wisdom and judgment they gain from accumulated experience before the Court.

Their expertise extends well beyond arguing the federal government's cases before the Court. Perhaps even more important, the solicitor general's attorneys are skilled at knowing which arguments and cases are losers for the federal government and how to preserve their lower court victories. To that end, when the United States has prevailed in the lower courts and the losing party seeks review, the Solicitor General's Office attorneys are experts at persuading the Justices to deny review. They know precisely what buttons to push, themes to promote, and concerns to exploit to convince a Justice—and the Justice's law clerks—that the case is not suitable for Supreme Court review.

A Justice's law clerks—recent law school graduates—are especially vulnerable to an effective argument by the Solicitor General's Office in favor of denying a cert petition. A law clerk's nightmare is to recommend to a Justice that a case be granted review and then, relying on that recommendation, the Court agrees to hear the case, only later to conclude after reading full briefing and hearing oral arguments that the case is not so important and should have been passed over. The Justices will feel they've wasted their time, and the clerk whose memo failed to spot the problems with the case is hugely embarrassed. So it's easier and safer to say no than to recommend that a case be heard.

The Solicitor General's Office is highly adept at exploiting these law clerk insecurities, partly because many of its attorneys harbored these same insecurities when *they* served as Supreme Court clerks. They know the cert process and have mastered the many ways to pitch a case to make a lower court ruling look far less important than it might have seemed upon first reading.

The solicitor general's briefs in opposition to a cert petition, referred to as "opps," rely on a menu of tried-and-true tactics. Sometimes cert opps will point to factual peculiarities in a case and argue that the ruling of the lower court is very much grounded in these "unique facts" and therefore of limited precedential significance. In other cases the cert opp will assert that the decision of the lower court is riddled with "procedural irregularities and obstacles" that would likely encumber the Court's ability to consider what might have at first glance appeared to have been an interesting, important legal issue. The Court is well advised to deny the petition in such instances, because the current case provides such a "poor vehicle" for the Court's review.[3]

In still other circumstances, the solicitor general cert opps will argue that even though the lower courts are in disagreement, there has been a recent "intervening development" that makes it more sensible for the Court to wait and see how the law develops in future cases. In short, the Solicitor General's Office knows all the advocate's tricks for persuading the Justices and their law clerks not to disturb a case that the United States has won in the lower courts.[4]

A Surprising Blunder

In the *Massachusetts* case, for some reason, the Solicitor General's Office uncharacteristically stumbled. The government's cert opp brief did not take full advantage of the many readily available grounds for persuading the Court that the case was clearly not worthy of its review. Worse still, the cert opp committed a highly unusual blunder that made the Court's review more likely, rather than less.

An effective brief opposing Supreme Court review has one overriding goal: to demonstrate that no important legal issue has been decided by the lower court and none is presented now for the Justices to review. Everything in the opp is written to make the case seem boring and inconsequential. An outstanding cert opp will be stripped of engaging adjectives or adverbs, or sweeping generalizations. While the cert petition seeks to make a case seem important and exciting (even if it is not), the opp strives to make it seem dry and irrelevant.

It should have been easy for the solicitor general to make *Massachusetts* appear inconsequential, as the D.C. Circuit ruling offered no majority ruling on any meaningful legal issue. The solicitor general's opp in *Massachusetts* started out well. It pointed out that a lower court judgment supported by separate opinions is "a particularly poor vehicle" for Supreme Court review. It stressed that there was a threshold jurisdictional hurdle in the case—whether the petitioners were suffering an injury of a type that was sufficient to give them the "standing" needed to bring the case—that might prevent the Court from ever reaching any of the legal issues presented by the petition.[5] But then, inexplicably, the government made a remarkably incautious statement. The solicitor general's opp defended the lower court's ruling by saying that because the relevant section of the Clean Air Act authorized the EPA to exercise its "judgment" in deciding whether greenhouse gases endanger public health and welfare, Congress had deliberately chosen to "not in any way cabin the Agency's discretion—procedural or substantive—to decide how to make that judgment most effectively."[6]

The last thing a lawyer should do in opposing Supreme Court review is to defend the lower court judgment based on a broad, sweeping proposition of law. By doing so, one makes the case seem far more, rather than less, important. Yet that is precisely what the solicitor general did with his brief. He had asserted a broad legal theory as a defense of the lower court by suggesting that the EPA's discretionary authority had not been limited "in any way."

When the *Massachusetts* petitioners received the solicitor general's brief, they could hardly believe their good fortune. They immediately pounced, fully exploiting the federal government's misstep. Petitioners in the Supreme Court are allowed to file a reply brief, and the *Massachusetts* petitioners' reply brief grabbed hold of that one errant sentence and went to town. The reply began with a heading in bold typeface and initial capitalization: "The Solicitor General's Assertion That EPA Has Virtually Unbridled Discretion Underscores the Need for This Court's Review."[7]

The brief was sharp and pointed: "On behalf of EPA, the Solicitor General invokes a radical vision of administrative law under which agencies can exercise unlimited legislative judgment immune from judicial review." In an effort to underscore that the case now involved far more than just the meaning of one provision in one federal statute, the brief pointed out that there are "literally hundreds of federal statutes that authorize agency officials to use their 'judgment'" and therefore the government's legal theory had "sweeping ramifications for executive branch authority."[8]

It was still far from a done deal that the Court would grant review. However much the *Massachusetts* petitioners sought to distract the Court with the "radical" arguments of the solicitor general, the odds were still stacked heavily against them because of the absence of a clear ruling by the lower court. But the odds had markedly improved.

"The Petition for a Writ of Certiorari Is . . ."

As soon as the *Massachusetts* petitioners filed their reply brief, all nine Justices began in earnest, with the help of their chambers, to review the briefs in order to decide whether review should be granted. By 2006 the Justices no longer had to wait for the clerk's office to deliver hard copies of the briefs. They could read electronic copies of the filings that were instantaneously transmitted to each chamber.

Not all nine chambers paid immediate attention to the *Massachusetts* cert petition. In 2006, eight of the nine chambers—all but Justice John Paul Stevens—participated in what they called the "cert pool" (created by Chief Justice Warren Burger in 1973) under which only one clerk working on behalf of all the eight chambers prepared a detailed memorandum on whether review should be granted. The memo was then circulated to the seven other participating chambers. By combining their resources in this manner, all eight chambers didn't have to prepare memos on each of the approximately ten thousand petitions for review filed every year with the Court. Each chamber was instead principally responsible for drafting the memos for one-eighth of all petitions, allowing the clerks to spend substantially more time writing each memo.[9]

Pool memos were supposed to be sharply focused on the bottom line—whether review was warranted. They should never assume any particular ideology, given that their audience would not just be the clerk's own Justice but Justices of widely varying viewpoints. Nor should the cert pool memo be especially long, because Justices do not have the time to spend hours on end deciding whether to grant review. Many memos were three to five pages long, though a memo might, in a particularly complicated matter, be as long as ten to twenty pages.

Although the cert pool memos included the clerk's recommendation, each Justice had his or her own law clerk review and "mark up" the pool memo and independently advise the Justice as to whether the clerk agreed or disagreed with its recommendation. Still, doing a markup took a lot less time than writing the memo in the first place.[10]

The clerks in Justice Stevens's chambers did not have the advantage of the cert pool, because Stevens, alone, chose not to participate in the system. He felt that it was important for his chamber to do its own independent work, rather than rely on the analysis prepared by others. He worried that the pool had a dampening effect on the Court's willingness to grant review. As he later explained on C-SPAN, the author of the pool memo, "when in doubt," does not recommend review because "you don't stick your neck out."[11] This, the Justice believed, had led the Court to grant too few cases in recent years. Stevens also believed that his chamber was more efficient at processing cert petitions than those that relied on the pool, which wasted a lot of time on petitions that clearly warranted denial.[12]

The Justices consider cert petitions on "conference days," scheduled several times each month (except during the summer recess), typically on a Thursday or Friday. They regularly review 150 to 250 cert petitions at each conference. Because the Justices cannot possibly discuss all 150, let alone 250, petitions in the two to three hours devoted to a conference meeting, the Chief Justice circulates to all chambers a few days before a scheduled conference a "discuss list" of those cases that, based on his chamber's review of the pool memo, the Chief believes worthy of possible review. Other Justices can then add to the discuss list other cases they believe might warrant review.[13]

Of the hundreds of cert petitions up for consideration at each conference, only about ten to fifteen make the discuss list. When the Justices convened for conference on June 15, 2006, the day scheduled for consideration of the *Massachusetts* cert petition, each Justice had on the conference table by their seat a list of some 250 cases with only the cases on the discuss list highlighted in yellow. All cases not highlighted would be denied review without discussion.

The day before the conference, something unusual happened. The *Washington Post* published an editorial, entitled "Warming at the Court," that spoke directly to the Justices considering the very next day whether to grant review in *Massachusetts v. EPA*.[14] What made the *Post*'s impeccable timing all the more remarkable was that the newspaper had not previously published any significant articles reflecting awareness of the litigation—not when the D.C. Circuit had ruled in the case the previous July, nor when it had denied a rehearing the past December, nor when the petition in the case was filed with the Court in March. Yet somehow the *Post* editorial staff knew precisely when the Justices would be considering the case in its private conference. To be fair, that date can be determined through publicly available channels, but it remains the kind of fact that only someone who is truly knowledgeable about the Court's internal operations would do the necessary legwork to discover, and then exploit.

The *Post* editorial was not just savvy in its timing. It was strikingly sophisticated in its content and had an unusually keen sense of how best to persuade the Justices to grant review. The piece freely acknowledged that because the D.C. Circuit's three-judge panel had "split three ways," the court's ruling for EPA lacked "coherent reason" and therefore was "now something of a dud." But then the *Post* argued that the Justices should nonetheless grant review.[15]

First, it made a case for relevance: the editorial stated that *Massachusetts* "could become one of the most significant pieces of environmental litigation in a generation." It then carefully identified and parsed the distinct legal issues raised by the case and discussed why, given the importance of the global warming issue, it made "no sense for the Justices to leave in place" the D.C. Circuit's "muddled decision."[16]

The day after the editorial's publication, the Justices postponed consideration of the *Massachusetts* petition for one week, from Thursday, June 15, until Thursday, June 22. Such a postponement was highly unusual then and most likely meant that some Justices supported review but there were not enough votes—four were needed—to grant review. Likely bolstered by the *Post* endorsement of the significance of the case, those Justices favoring review secured an extra week to lobby their colleagues. So what led the *Post* to publish such a timely and smartly directed editorial? Who had engineered the piece?

The author of the *Post* editorial was plainly Ben Wittes, who served at the time as the newspaper's legal affairs editorial writer and had a deserved reputation as an expert on both the D.C. Circuit Court and the Supreme Court. Wittes spoke fairly frequently with several D.C. Circuit judges to learn about pending and upcoming matters of possible interest to *Post* readers. During one of those conversations, Judge Tatel had told him about the D.C. Circuit *Massachusetts* ruling, outlining the legal issues raised in the petition pending before the Supreme Court and stressing their importance. Based on what he had learned, Wittes must have decided the case was worthy of an editorial.[17]

The *Massachusetts* petitioners had no advance notice. The Sierra Club's Bookbinder spotted the *Post* editorial first early in the morning and quickly dashed off an email to co-counsel, including Milkey. Milkey took only two minutes before forwarding it to Heinzerling. He was both pleased and mystified about its origins. Clearly they had a powerful friend somewhere, who was trying to give them a much needed, last-second boost across the finish line. Perhaps, after all, something miraculous might happen.[18]

On the morning of Monday, June 26, dozens of attorneys across the country who had filed briefs both for and against Supreme Court review anxiously awaited the Court's order list from its conference on Thursday. All had expected a routine denial a week earlier, but the surprise one-week delay suggested something highly unusual might be up. The list would be posted online at 10 a.m.

By midmorning, the attorneys from the environmental public interest groups, state attorneys general's offices, and industry were all intently focused on the clock. The EPA career attorneys were eagerly awaiting zero hour, some privately hoping that the Court would grant review and reverse their own agency's decision. Joe Mendelson, whose petition had started the whole thing more than six years earlier, waited impatiently at his computer. Perhaps Judge Tatel did too.

At precisely 10 a.m. Mendelson clicked on the Supreme Court's website and scrolled down the Court's list of cases granted and denied review that morning. In 238 cases, the Court had issued its routine statement: "The petition for a writ of certiorari is denied." He nervously scanned down the order list, looking for the *Massachusetts* case. Then he saw the listing for Case No. 05-1120, *Massachusetts v. EPA*. "The petition for a writ of certiorari is granted."[19]

Mendelson was floored. "Holy #@$#!," he wrote in an email.[20] In Boston, Milkey's reaction was similar but more direct: "Holy Shit."[21] Back in D.C., Sierra Club's Bookbinder's was slightly more secular. "Oh, shit," he remembered thinking. "Now what?"[22] The Justices would hear their first-ever climate case.

"What have we done?" Milkey thought to himself, becoming increasingly unsettled as the enormity of the challenge sunk in. He later recalled that he "felt a disconnect because everyone else, including all the people who did not want us to keep the case alive, were elated, and I was severely anxious." Persuading the Justices to hear a case was not the same thing as getting them to decide a case in your favor.[23] And what if Doniger was right? What if the Justices ended up closing the door on future climate litigation and he was responsible for setting back environmental regulation for a decade or more? Had he been wrong not to wait?

As hard as it is to convince the Court to grant review, it requires the votes of only four Justices to grant a cert petition. To win, the *Massachusetts* petitioners would now need five. "The future of the environmental movement" might truly now be, Milkey contemplated, squarely on his "head"—just as NRDC's Frances Beinecke had warned.[24]

12

The Lure of the Lectern

When Lisa Heinzerling accepted Jim Milkey's invitation to serve as the principal brief writer, they agreed to postpone deciding which one of them would present oral argument in the Supreme Court. Unlike the D.C. Circuit, the Supreme Court almost never allows oral argument by multiple lawyers on the same side. The Court rules are quite clear: "Only one attorney will be heard for each side"—absent highly unusual circumstances.[1] The rule does not bend even when, as in *Massachusetts,* there are dozens of parties on that same side.

In Supreme Court litigation, emotions run high. The Court's insistence on only one advocate per side can lead to pitched battles over which lawyer gets the nod. The stakes are huge, potentially historic. The petitioners all naturally agree that the best lawyer should do the argument, but they often vehemently disagree about who that best lawyer might be.

For most lawyers, presenting an argument before the Supreme Court would fulfill their wildest professional dreams. But only an infinitesimally small fraction of the more than 1.3 million lawyers in the United States ever gets that opportunity. Not many more than one hundred, and sometimes fewer, argue before the Justices each year, and the vast majority of those are experienced Supreme Court advocates with multiple arguments under their belt. There are very few opportunities for rookies to play in the major leagues of the Supreme Court.[2]

For that same reason, a lawyer who argues a Supreme Court case gains a unique professional credential that can lead to a string of career opportunities. A single successful Supreme Court argument in a high-profile case can be a complete game-changer in one's career. It is no exaggeration that there are federal court of appeals judges and even Supreme Court Justices whose appointments can be traced to the celebrity they first enjoyed as a result of a successful Supreme Court argument. That is why it is very hard for any lawyer to give up an opportunity to argue before the Court, especially when it is unlikely that another chance will ever arise.

In some instances the Court's formal intervention has been necessary to overcome a disagreement over who should present oral argument. In several cases, competing lawyers have resorted to flipping a coin. When, in one case, the coin flip was won by a novice advocate who insisted on doing the argument instead of the experienced Supreme Court advocate who lost the toss, the novice's argument was so disastrous that it led to a malpractice claim against the oral advocate.[3]

When the Supreme Court formally granted review in *Massachusetts v. EPA* on June 26, 2006, the petitioners quickly reached a consensus that Lisa Heinzerling should be their principal brief writer. The decision was an easy one. Heinzerling had rescued the case by drafting the outstanding cert petition that had persuaded the Court to grant review. But they deliberately declined to decide in June who would present their oral argument. They reasoned that they could postpone that decision until the end of August, when their legal brief was due, as there would be ample time then to make a decision—the oral argument would occur in late fall at the earliest. Making the decision immediately after the Court had granted review would distract everyone from their highest and most immediate priority, which was the filing of an outstanding legal brief. They all knew they could not win their case absent such a brief.

Left largely unstated was that there was likely to be a fight over who would do the oral argument. It was clear to all that both Milkey and Heinzerling hoped to be the oral advocate, and a handful of others on the team likely privately hoped that the fates might somehow conspire for them to receive the coveted assignment. But rather than face that difficult question in June, they all decided to kick that can down the road. The postponement placed a time bomb within the petitioners' legal team that, when later triggered, irreparably destroyed professional relationships and came close to dooming their case.

The Famous Carte Blanche

When Heinzerling agreed in late June to draft the petitioners' written brief, she made only two requests of Milkey. First, she wanted to be formally listed in a prominent position on the brief as a "Special Assistant Attorney General" for Massachusetts.[4] It was a reasonable request—reflecting her status as the brief's principal author. Second, she wanted to be seriously considered for the oral argument. She understood that Milkey was also interested in that role and agreed that the decision could be postponed until after the brief was filed. Milkey agreed to both conditions.[5]

Once she was formally on board, Heinzerling faced an August 29 filing deadline and a daunting logistical challenge. Because the petitioners were all filing one joint brief, her work needed to be reviewed and approved by twenty-eight parties and their approximately sixty lawyers. Satisfying their disparate views would be time-consuming. Worse, it could result in the kind of fragmented brief-writing process that had undermined the quality of the brief that they had filed in the D.C. Circuit. No one had paid much attention to the cert petition because they had not expected it to be granted. This time, everyone wanted a piece of the pie. Heinzerling would have to walk a tightrope. She needed to consult with her numerous co-counsel—many of whom would have valuable ideas and insights—and obtain their buy-in. But she could not afford to cede ownership of the brief's content, flow, and voice.

Heinzerling circulated draft briefs on July 28, August 18, and August 25, and filed a legal brief with the Court for all twenty-eight petitioners on the August 29 deadline. The final, outstanding brief reflected some bold and risky strategic calls. And it very much reflected her distinct personality and engaging writing style.

Beginning with the initial draft, Heinzerling decided to reverse the order and emphasis of the legal issues from how they had been presented in the cert petition. While the cert petition had been designed to persuade the Court to grant review, the brief now had to convince the Court to rule in favor of the *Massachusetts* petitioners. The cert petition had presented the legal issues in the order most likely to interest the Justices, focusing first on the EPA's backup argument that even if greenhouse gases were air pollutants, the agency had the discretion to postpone its decision on whether greenhouse gas emissions from new motor vehicles endangered public health and welfare. As dressed up by Heinzerling's cert petition, the first legal issue had been smartly recharacterized as involving the sweeping importance of the power of an executive branch agency (the EPA) to substitute its view of sound policy for that of Congress. The environmental dimension of the case had been deliberately de-emphasized.[6]

The cert petition's "second issue"—whether greenhouse gases were "air pollutants" subject to the EPA's regulatory authority under the Clean Air Act—was of course the legal issue about which the petitioners cared the most.[7] But the petition did little more than raise that issue without any elaborate discussion: because the lower court had not ruled on the issue, Heinzerling had known it would be hard to convince the Justices that they should do so now.

Now that review had been granted, Heinzerling's brief quickly discarded any pretense of real interest in what, not long before, had been that all-important "first issue." There was nothing underhanded about the shift. Both the cert petition and the merits brief contained an initial "Questions Presented" section, listing in order the legal questions to be considered by the Court. Between the petition and the brief, Heinzerling had simply reversed the order of those two issues.[8]

To that same end, the "Argument" section of the brief focused almost exclusively on the greenhouse gas issue and gave short shrift to the question of the power of an executive branch agency to substitute its views for those of Congress. Indeed, of the fifty-seven pages in Heinzerling's first draft of the brief, only six directly addressed this issue, previously litigated as the case's main draw.[9]

Such "bait and switch" tactics are regularly undertaken by experienced Supreme Court advocates. They know that the legal issues and arguments that are the most helpful for persuading the Court to grant review are not necessarily the ones that will convince the Court to rule in their client's favor. Heinzerling's gambit, particularly her formal reversal of the legal issues in the "Questions Presented," was an especially gutsy ploy because it was about as subtle as a punk rock band playing in a train's quiet car. Even more essential to the effectiveness of the brief was Heinzerling's decision to break free of her reviewers when the brief was otherwise at risk of dying a death by a thousand edits.

The initial draft, circulated on July 28, was a very good first cut, given the speed of its production. But it was no more than that: a first draft. And, like any first draft, it was invariably rushed in its analysis in some parts. Many co-counsel weighed in by email with suggested line edits, and a smaller group participated in a lengthy multi-hour conference call in which they went through the draft together line by line until, as Heinzerling described it in a follow-up memo, they all "ran out of time and steam." The main contributors were Mendelson, Bookbinder, Doniger, Fox, and a bevy of attorneys from the attorneys general offices of California, New York, and Massachusetts, including Milkey. Everyone had their own ideas and suggestions. Mendelson's comments were among the quickest and the most detailed, and were very constructive by offering specific suggestions for word changes. Milkey's comments were potentially the most distressing to Heinzerling because he raised a series of "big picture concerns" about significant portions of the brief without much specific guidance on how these concerns could be addressed. The sheer number and variety of comments made clear that herding these cats to a consensus would likely be a difficult, time-consuming, and tiring process.[10]

Although the second draft, circulated on August 18, included good ideas and had benefited in substance from the input of many knowledgeable reviewers, it was in some significant respects less effective than the first. Substantial portions of the revised draft read like a brief written by a committee, and it was weighed down by Heinzerling's efforts to accommodate the enormous number of competing comments. With only eleven days remaining before the filing deadline, Heinzerling warned her reviewers that her "head will explode" if they continued to offer line-by-line edits in the final days before the filing deadline.[11]

Six days later, while continuing to acknowledge the "excellent comments and suggestions" she had received, Heinzerling went rogue. She circulated a significantly revised draft in which she no longer tried to satisfy all the individual requests she had received. She instead crafted new arguments with more depth and verve. When she sent out the third draft to her reviewers, Heinzerling explicitly warned them about what she had done so they would not be "shocked" when they read the latest draft, which was "very different from the last." She urged them "to read it with a fresh and open mind" and strongly intimated that everyone would have to cease the endless wordsmithing to permit the brief to be filed on time.[12]

The revised draft hit all the right chords. The responsive praise from co-counsel was universal. The California Attorney General's Office and Milkey both responded identically—"thank you, thank you, thank you"—and remarked on the high quality of the new draft.[13] NRDC's David Doniger proclaimed the draft "terrific" and others labeled it "fabulous."[14]

Not only was the latest draft's legal argument far more focused, direct, and forceful, but the prose was confident, engaging, and entertaining. "Jaunty" is how Heinzerling herself described it, allowing the brief to have "spring in its step."[15]

The brief cheekily characterized the EPA as engaging in "a host of interpretive don'ts: it ignored statutory language, inverted the usual meaning of other language, interpreted the same words to mean different things, and shrugged off Congress's explicit determination that an effect on climate is an important component of human welfare." The brief's evocative choice of words enhanced its argument: "Making up one's mind first and then looking for reasons to support one's decision is the very *soul* of arbitrariness."[16]

The closing words of the final brief filed with the Court on August 29 were the brief writer's equivalent of an Olympic figure skater landing cleanly on the ice following a triple axel. It derided the opinion of the D.C. Circuit's Judge Randolph:

> Indeed, the approach taken in the lead opinion below mocks the very process of judicial review. The purpose of judicial review is not well served when courts approve agency action with reasoning that sounds like Alexandre Dumas's famous *carte blanche*: "It is by my order and for the good of the state that the bearer of this has done what he has done."[17]

Heinzerling's original draft of that language, however, had been slightly different and, many thought, better. She had originally referred not to Dumas himself but, more accurately, to his depiction of Cardinal Richelieu in *The Three Musketeers,* who (in the novel) wrote a letter granting its bearer *carte blanche* authority.[18] The last-minute switch from quoting the cardinal to quoting Dumas was made only hours before the final brief went to press, in response to Milkey's lone concern that some of the (Catholic) Justices might view the Richelieu reference as somehow "Anti-Catholic."[19] The final brief, with the cardinal left on the editing floor, was filed. It was a small and seemingly innocuous edit, but it pointed to tensions just below the surface that would very soon boil over.

"A Beauty Contest"

Almost as soon as the brief was filed, the *Massachusetts* petitioners knew they could no longer delay deciding who would present their oral argument. The choice was legitimately hard: a good argument could be made for either Milkey or Heinzerling.

Milkey, after all, was the petitioners' official "Counsel of Record," reflecting the central role that he had played in the case. He had championed the decision to seek further review when many of the rest of them had opposed him. Although Milkey had never argued a case before the Supreme Court, he had substantial experience as an appellate advocate, including in both federal and state appellate courts. Milkey also had an impressive academic and professional record of achievement and had devoted his twenty-year career to public service. To the extent that the petitioners had agreed that having a state as the lead petitioner bolstered their prospects before the Justices, there were related strategic advantages to choosing a career public servant from that state. Milkey would stand before the Justices neither as a "hired gun" nor as an elected official seeking the limelight to promote his political future. He could project the image of a public servant who had dedicated his entire career to safe-guarding the interests of his state and its citizens.[20]

But there were also good reasons to give Heinzerling the nod. More than anyone else on the petitioners' legal team, she was responsible for the outstanding quality of their successful cert petition and their brief. Twice, she had delivered top-notch written briefs in challenging circumstances. Heinzerling was clearly a brilliant lawyer and a gifted writer. She had mastered the legal issues and arguments as well as anyone on the team, even though she had not joined the team until late in the game. Heinzerling was also quick and effective on her feet, a much-needed skill when answering the Justices' rapid-fire questions during oral argument.

That said, she had never before argued in any appellate court, federal or state. An argument in the U.S. Supreme Court in *Massachusetts* would be her first appellate argument ever—certainly a high-stakes novice run. But she had a reputation as an outstanding classroom teacher and as an engaging legal scholar—all skills that were not unrelated to those of an effective appellate advocate. As a former Supreme Court clerk, Heinzerling also had an insider's invaluable perspective on the Court.

The conflict over oral argument came to a head in mid-September, just weeks after the brief's filing, by which time Milkey had concluded that his courtroom experience and status as a career public servant made him the better candidate. NRDC's Doniger and Sierra Club's Bookbinder, however, had decided that Heinzerling should be the oral advocate. Neither told Milkey of their view. Instead, they worked behind the scenes to develop a game plan for persuading the other petitioners to choose Heinzerling by showcasing her superior speaking abilities. They would have her participate in several high-profile public presentations about the Supreme Court case during the month of September and then tout her outstanding performance to the other co-counsel.[21]

Milkey would hardly have been surprised to learn that Doniger opposed his doing the oral argument. The two had been like oil and water from the outset of the case. Doniger had been dismissive of

Milkey's writing in the D.C. Circuit brief and of his comments on Heinzerling's draft Supreme Court brief. Doniger later said that he had never "trusted" or "agreed with [Milkey's] judgment." By contrast, Doniger had been very favorably impressed by the high quality of Heinzerling's work.[22]

But Bookbinder's early opposition would have surprised Milkey, as he had frequently left Milkey with the impression that he strongly favored Milkey for the argument. Bookbinder had been put off by Milkey's comments on Heinzerling's draft briefs. After meeting with Heinzerling and becoming aware of the unhappiness that Milkey's comments were causing her, he even flew up to Boston to tell his former boss that his thinking on the case bordered on the "irrational." Bookbinder worried at one point during the drafting process that Heinzerling was so distressed she might leave the case.[23]

Doniger and Bookbinder both knew it would not be easy to persuade the rest of the petitioners' legal team, all of whom were practicing lawyers, to favor an academic with almost no courtroom experience over an experienced advocate like Milkey. The other state attorneys general offices, in particular, would be naturally loyal to Milkey, who was one of their own.

Peter Lehner, chief of the New York Attorney General's Environmental Protection Bureau and a former colleague of Doniger's at NRDC, contacted Milkey in late September to notify him that "some people" thought that the question of who should present the Supreme Court oral argument should be decided by an internal contest.[24] As described by Lehner, both Heinzerling and Milkey would participate in a formal competition—sometimes referred to as a "beauty contest" or a "bake-off"—in which they would each present an oral argument before a panel of attorneys, who would ask questions as though they were the Justices. Whoever did the best job would be chosen to present the argument before the actual Justices. Heinzerling had already agreed to participate in the competition.[25]

Milkey heard Lehner out, and flatly rejected the idea.

He argued that any such competition would be based on negligible to no meaningful preparation and therefore "would bear very little resemblance to an actual courtroom experience."[26] It would be wastefully counterproductive, Milkey argued, whether done immediately or after the opposing briefs were filed. If done now, they wouldn't even know the other side's arguments. (The opposing briefs were not due to be filed for four weeks.) And if done later, the time they would need to prepare their reply brief would instead be spent on conflicts.

The next day Bookbinder wrote to Milkey, asking him to reconsider "for the benefit of what is the most important environmental case in the country."[27] Milkey rejected Bookbinder's request. He said he had surveyed a number of "disinterested outsiders" for their views and "they were unanimous in their opinion that I should do it, and that this was not even a close call."[28] Milkey added that "two people commented that—no matter how smart Lisa is—sending someone to the Supreme Court for her first appellate argument of any kind was tantamount to malpractice."[29]

For Milkey, the dispute had become highly personal. As he reminded Bookbinder, the two of them had long ago privately agreed that having Heinzerling write the brief and Milkey argue the case was the "obvious" solution.[30] According to Milkey, that division of responsibility had been Bookbinder's own suggestion. Milkey thought Bookbinder was letting his almost three-decade friendship with Heinzerling, beginning as Princeton undergrads, cloud his judgment. Milkey also told Bookbinder that he blamed Doniger for creating the mess in the first place: by promoting the "competitive moot idea" and "egging Lisa on" because of his "Ahab-like obsession" with and dislike for him that had led Doniger, again and again, to undercut Milkey's role in the case.[31]

The interwoven threads—what was in the best interest of the case and what was in the best interest of individual lawyers—long present but largely submerged, were now on full display. And fraying.

The dispute was abruptly settled a few days later, on October 4, when Milkey's immediate supervisor within the Massachusetts Attorney General's Office made it clear that because Heinzerling served as a Special Assistant to the Massachusetts Attorney General, Massachusetts Attorney General Tom Reilly had exclusive authority to decide whether she could present oral argument.[32] Neither Doniger, nor Bookbinder, nor any of the other co-counsel could dictate to the contrary. It was Reilly's decision, and he was choosing Milkey.

Nor could Reilly's authority be evaded by switching Heinzerling from counsel for Massachusetts to counsel for any one of the other dozens of petitioners. Apparently under Massachusetts ethics law, absent the attorney general's express permission, Heinzerling could not now switch clients and become counsel to one of the other petitioners. And no such permission would be given.

That afternoon Heinzerling sent an email to the entire litigation team, formally withdrawing from consideration for the oral argument. She defended her credentials for doing the argument, expressed her continued belief that she "could do not only an excellent job, but the best job," and restated that, unlike Milkey, she had been willing to do a "'competitive moot' to sort out the choice about oral argument."[33] However, "the uncertainty has gone on too long" and its "prolonged and unpleasant nature . . . threatens the case."[34] She later acknowledged that some of those who had wanted her to do the argument were "aghast" at her decision to withdraw.[35]

Anyone in Heinzerling's position would have been understandably upset. She had extraordinary academic credentials, had worked hard on the case, and had produced excellent work, only to be labeled too inexperienced by someone whom she had long considered a mentor and close friend. She had also lost an important professional opportunity. (A Georgetown faculty colleague and contemporary who'd had the good fortune a few years earlier to argue and win a high-profile civil rights case before the Court later became a judge on the D.C. Circuit partly due to the celebrity resulting from her successful Supreme Court advocacy.)[36]

In the aftermath of such an enormous disappointment, many might have walked away from the case, taking the position that if Milkey was doing the argument, he should also do the reply brief that would be due in a few weeks. To her credit, Heinzerling did not. Her commitment to the cause proved stronger than any individual ambition. She agreed to stay and take on the reply brief later that month. She had only one condition: her co-counsel could not insist, as they had before, on being able to comment and propose edits on every sentence and word. For the good of the brief and for her own peace of mind, she was "not going to both stand down on the oral argument and take a four hour line by line conference call."[37] She would do the reply brief only if the many co-counsel would be limited in their comments.

If Heinzerling had left the writing of the reply brief to others, it could have crippled the petitioners' ability to file an effective reply brief. Milkey would have been buried in oral argument prep and none of their other co-counsel had displayed her talent for writing. Instead, she swallowed her pride and agreed to play a subordinate role in the most high profile part of the case, while doing her part to make the case as strong as possible.

The Opposition Fights Back

On October 24 the solicitor general filed a fifty-page brief on behalf of the EPA that responded to the *Massachusetts* petitioners' brief. Unlike the Justice Department's brief in the D.C. Circuit, which suffered from political appointee rewriting, this brief was excellent.

It began with ten pages of hard-hitting arguments that the Court should dismiss the case because the *Massachusetts* petitioners lacked "standing" as required by Article III of the Constitution. According to the brief, the petitioners' standing argument fell short in two respects. First, they had failed to establish that the EPA's denial of Mendelson's petition would "cause" them an imminent, concrete injury. And second, even if the EPA's action would cause such injury, the petitioners had failed to establish that a court could effectively "redress" that injury.[38]

The gist of the solicitor general's argument was that the physics of climate change—that greenhouse gases are emitted by sources all over the globe and evenly concentrated in the atmosphere over decades—rendered too "speculative" the question of what specific sources were causing which climate change impacts across the globe. The *Massachusetts* petitioners therefore could not show how the EPA's failure to regulate motor vehicle emissions, as distinct from all those other sources, was causing harmful climate change.[39]

The solicitor general's second argument—that the petitioners had failed to demonstrate that any judicial relief the Court could provide in this case would "redress" their climate injury—also relied on the science of climate change. The solicitor general pointed out how small a fraction of total global greenhouse emissions was implicated by this litigation. Millions of sources around the globe had been for decades pouring billions of tons of greenhouse gases into the atmosphere and would continue to do so regardless of the outcome in this case. "Nothing in the record," the solicitor general's brief concluded, "suggests that so small a fraction of worldwide greenhouse gas emissions" as those generated by new motor vehicles in the United States—the only emissions covered by Mendelson's petition—"could materially affect the overall extent of global climate change."[40]

Like the causation argument, the redress argument had real force. It would likely persuade some Justices that the *Massachusetts* petitioners lacked standing to bring the lawsuit. The question was how many.

The solicitor general's treatment of the question of whether greenhouse gases constituted "air pollutants" subject to the EPA's Clean Air Act regulatory authority was also much stronger than the earlier

D.C. Circuit brief had been. The solicitor general wisely abandoned the arguments that the political appointees in the Justice Department and the EPA had previously insisted on. In particular, the solicitor general did not argue that the EPA's position was compelled by the plain meaning of the Act—the argument that the political appointees had favored because it would prevent a future (Democratic) White House from using the Clean Air Act to regulate greenhouse gases.[41]

The solicitor general instead embraced the far easier, and far stronger, argument that the EPA's interpretation was "reasonable" and therefore "entitled to deference" under the Supreme Court's 1984 unanimous ruling in *Chevron v. Natural Resources Defense Council*, written by Justice John Paul Stevens. To prevail on this less demanding standard, the solicitor general needed only to persuade the Justices that, as evidenced by the Clean Air Act's language, congressional intent to regulate greenhouse gases as air pollutants was at least ambiguous and that the EPA's interpretation of that ambiguous language was not unreasonable.[42]

In support of this argument, the government had two good points. First, Congress had not specifically contemplated regulation of greenhouse gases when it passed the Clean Air Act in 1970. That basic fact was largely undisputed. And second, the Clean Air Act assumed characteristics about air pollution—for instance, that pollution concentrations in the air varied by geographic location—that were not true for greenhouse gases, for which there is one uniform atmospheric concentration across the entire globe. As a result, some of the Clean Air Act provisions would make little sense if the term "air pollutant" were defined to include greenhouse gases.[43]

The solicitor general's brief similarly improved upon the briefing of the Justice Department in the D.C. Circuit on the reasons the EPA's decision to deny Mendelson's petition was lawful even if greenhouse gases were Clean Air Act pollutants—the agency's backup argument for winning the case. Both the EPA's original decision and the Justice Department's D.C. Circuit brief had been unnecessarily confusing on this issue. Neither made it sufficiently clear what EPA "decision" the government was defending. Were they defending a decision that greenhouse gases from motor vehicles do not endanger public health or welfare? Or were they defending a decision not to decide whether such an endangerment exists?

Here too, the solicitor general chose the easier and more defensible approach. Unlike in the lower court, the Supreme Court brief made it clear that the EPA had merely postponed its decision on whether an endangerment existed. The government's brief went on to defend the agency's postponement based on the settled principle of administrative law that "an agency's decision not to initiate a rulemaking may be based on a wide range of discretionary factors and is reviewed under a highly deferential standard."[44]

The brief further pointed out that, unlike other parts of the Clean Air Act, the specific provision of the Act upon which the *Massachusetts* petitioners relied did not divest the EPA of an agency's ordinary discretion to decline rulemaking at one time in favor of revisiting the issue at a later time. Specifically, the Clean Air Act neither imposed any deadlines for when the EPA must respond to a petition to regulate new motor vehicle emissions, nor expressly confined the factors the agency could consider in deciding that the time was not yet right.[45]

The solicitor general's brief, however, was not the only one filed by parties in support of the EPA. Four distinct groups intervened as parties in the case in support of the EPA: nine states, led by Michigan; the "CO_2 Litigation Group," representing many of the nation's leading trade associations and business organizations; the "Utility Air Regulatory Group," representing the nation's power plants and regulated utilities; and a group of trade associations representing the automobile, truck, and engine manufacturers, and automobile dealers. In the D.C. Circuit, the three industry groups had filed one joint brief.[46]

Unlike the *Massachusetts* petitioners, who had decided to file one joint brief to underscore their consensus, these four groups decided that it was to their strategic advantage in the Supreme Court to have each one file its own fifty-page brief, and to coordinate their filings.[47]

Two of their briefs exclusively addressed the question of whether greenhouse gases were air pollutants and largely ignored the legal issues of Article III standing and the EPA's discretion to deny the petition even if greenhouse gases were air pollutants.[48] A third brief devoted half of its argument to the standing issue.[49] And the fourth dedicated many of its pages to whether the EPA had discretion to decide not to make an endangerment determination.[50]

The decision to file four separate briefs cleverly took advantage of the Court's rules and had the effect of requiring the petitioners to respond to many more pages of arguments against them. The *Massachusetts* petitioners jointly filed only one 48-page brief. The solicitor general, industry groups, and states that were parties to the litigation responded with more than 200 pages in legal briefs.

The briefs of the petitioners and respondents were not the only briefs filed, however. They were joined by ten more briefs totaling several hundred pages filed by nonparty *amici curiae* (friends of the court) who all supported the EPA—economists, scientists, and business organizations—and fourteen *amici curiae* briefs also totaling more than several hundred pages filed in support of *Massachusetts,* by states, native tribal governments, mayors, former federal government officials, business interests, religious organizations, environmentalists, and scientists.[51]

The Reply

The petitioners' only opportunity to address this barrage of words was in a reply brief of no more than thirty pages (expanded by the Chief Justice in this case from the normal twenty-page limit). A reply

brief is always important in Supreme Court litigation because it allows the petitioners one final chance to answer the other side's arguments before oral argument. This is the petitioners' chance to ex-
pose the flaws in the opposing arguments and to rehabilitate their
own. It is the last brief that will be read by the Justices and their
clerks as they prepare for oral argument.

In the *Massachusetts* case, an effective reply brief was both particularly important and exceedingly difficult. Reading opponents'
briefs and those filed by their supporting *amici* in a Supreme Court
case is nerve-wracking. In *Massachusetts,* it was especially so. The
opposing briefs had been prepared by legions of the most highly
skilled Supreme Court lawyers in the country, both in the Solicitor
General's Office and in the nation's leading law firms representing
private industry. Because a case like *Massachusetts* implicated the
economic interests of major swaths of the nation's economy, leading
private sector Supreme Court lawyers could charge their business
clients top dollar: $800, $900, even $1,000 per hour.

Perhaps the opposing counsel will discover and highlight for the
Justices significant errors that the petitioners unwittingly made in
their own briefs. Or, precisely as was the case with the solicitor general's brief, sometimes the opposing briefs will make an entirely
new and far stronger legal argument. Supreme Court briefs are frequently orders of magnitude better than briefs filed in the same case
in the lower courts.

Yet, under the Court's rules, because the *Massachusetts* case was
set for oral argument on November 29, only four weeks after the
filing of the opposing *amicus* briefs, petitioners would have at most
three weeks to file their reply. That brief would need to—in relatively
few pages—respond effectively to all the opposing arguments made
in those hundreds of pages of briefs.[52]

Thanks to Heinzerling's insistence that her draft would be limited
in its review by co-counsel, the reply brief was filed on November 15.[53]
It dedicated little time to the issue about which the petitioners cared

the most—their argument that greenhouse gases are air pollutants— and that was for good reason. On this point, the opposing arguments by the EPA and the other respondents were fairly weak and had done little, if any, damage to the petitioners' position that greenhouse gases were clearly "air pollutants." Heinzerling could therefore use the reply brief's limited pages to address other, more important issues.

The reply spent more pages on the standing issue, which the petitioners had declined to address in their opening brief, than on any other issue. They could not afford to lose on standing. Their reply also sought to refute the EPA's reliance on "background principles of administrative law" for deciding not to reach the endangerment determination.[54] The brief stressed that the agency had not, in fact, relied on any of those principles when it declined Mendelson's petition in 2003. The EPA could not now justify its decision on new grounds.

The reply brief was good, but fell short of the exceptionally high quality of the petitioners' opening brief. That was unsurprising. By the time of the brief's filing, with only fourteen days remaining before the oral argument, many of the Carbon Dioxide Warriors were no longer on speaking terms. But at least the petitioners had managed to file a solid reply under challenging circumstances—with little time to respond to hundreds of pages of opposing briefs and a litigation team in disarray.

Only oral argument remained.

13

Moots

Jim Milkey was badly shaken. He was at home on a Saturday morning, the phone to his ear, speechless. It was November 18, 2006. In only eleven days, he would be presenting oral argument before the Supreme Court. The surprise caller, David Bookbinder of the Sierra Club, a good friend for more than a decade, had just asked him to step aside and let someone else argue the case. Other key petitioners no longer had faith in his ability to do the job, Bookbinder told him.[1]

As the words sank in, Milkey had to catch his breath. He felt blindsided. He was tired. He had been preparing for the oral argument full-time for the past six weeks. The last three of those weeks had been especially brutal. Only the day before, Milkey had completed the last of three demanding practice oral arguments designed to prepare him for facing the real Justices. He had thought the question of who would argue before the Court had been settled beyond dispute. Long ago. Bookbinder's call had proven him wrong.

As the *Massachusetts* petitioners' shared desire to force the EPA to regulate greenhouse gas emissions competed with pride and ambition, professional ties and personal friendships had frayed to the point of snapping. The two motivations were always interwoven—the common desire to ensure the best possible outcome and the individual desire to enjoy the prestige and recognition that came from arguing the case before the Supreme Court. Disentangling the two had become no easier as the Court date loomed.

Milkey's state of mind began to shift. At first he had been shocked, but now he grew angry. He strongly believed he had not let his ego interfere with his judgment that he was the best person on the team to deliver the oral argument. He "was never the kid in high school who was in the school play," he later recalled. Hell, he had never even wanted to be a litigator. Nor was he one of those people who dream of having a Supreme Court case. "That was never a driving force," he thought to himself. The decision had been made long ago that a state should be the public face of the case, and Massachusetts had been chosen for that pole position. Milkey was certain that, as both an experienced oral advocate and a lead counsel for Massachusetts, he was the right person for the job.[2]

After catching his breath, Milkey finally answered Bookbinder, who was waiting quietly on the other end of the phone. "That's not going to happen," Milkey coldly replied.[3]

"Be Prepared to Answer"

Having secured the oral argument in early October, Milkey had his work cut out for him. He knew from conversations with experienced Supreme Court advocates that preparing for the argument would require approximately 150 to 200 hours of his own time in addition to the work of his colleagues within the Massachusetts Attorney General's Office who would be helping him prepare. The Court schedules one hour for oral argument, evenly divided between the two opposing attorneys. Several hundred hours may seem like a lot of prep time for thirty minutes before the Justices. But it's needed.

In no more than thirty minutes standing at the lectern before the Justices, Milkey would have to both deliver any affirmative points he wanted to make and answer the barrage of fifty to seventy-five questions that the Justices would likely ask him. There would be no extensions of time, no time for missteps. He would have to make every second and every word count.

Expert Supreme Court advocates advised Milkey that to achieve the necessary high level of preparation, he would need to focus on three tasks. First, he would have to decide on the two or three most important legal arguments that he wanted to get across to the Justices. He had to be able to deliver these arguments in no more than five minutes, as that was the maximum amount of time he could anticipate not being interrupted with questions. His best shot at getting those few minutes of uninterrupted time was at the very outset, when the Justices typically (but not always) allow the advocates to give an opening statement before jumping in with questions. By putting his most important arguments on display for the Justices as early as possible, he could quickly discover whether they had concerns about the legal arguments most essential to his case.

Questions at oral argument are an advocate's friend. They reveal a Justice's thinking and provide an opportunity to address any concerns the Justice might have. Oral arguments fail when the advocate does not learn these concerns and, as a result, never has the chance to address them. There is nothing worse than losing a case on a ground that the attorney believes she could have responded to persuasively had she only known what the Justice was thinking.

Milkey's second task was to anticipate all the questions he might be asked and develop the best possible crisp and direct answers to those questions. An advocate never, ever wants to be surprised by a question during the actual argument. Anticipation and preparation are key.

As a result, advocates reread everything: all the briefs filed in the Supreme Court; all the prior, related decisions of the Court; and all the federal statutes, agency regulations, and constitutional provisions relied upon by the briefs filed with the Justices. The Court's official printed *Guide for Counsel in Cases to Be Argued before the Supreme Court of the United States* leaves no doubt as to the unforgiving expectations of the Justices. It admonishes counsel to "know the record" and "be prepared to answer" any question a Justice might have about it. The bar for advocates to clear is high: "It is impressive when you can respond with the volume and page where the information is located." "It is also quite effective to quote from the joint appendix," the guidance continues; that appendix can itself be many hundreds and sometimes thousands of pages long.[4]

As he combed through all that material, Milkey had to think of the hardest, most damaging questions he might be asked. He had to come to grips with the weakest, most vulnerable parts of his side's legal arguments. And he had to do so without getting discouraged. The advocate who is so convinced of the correctness of his argument that he fails to anticipate weaknesses and limits in that argument is a poor advocate.

A Supreme Court oral advocate has to become a master of more than just the legal arguments. He has to master the relevant facts— including, in this case, the physics and chemistry of climate science. The Justices are smart, curious people, and they want to know how the legal arguments apply to the real world.

Here again, the Court's official guidance to counsel makes clear the Justices' high expectations. The guide touts as an example of outstanding advocacy an attorney who was able to answer a Justice's question: "What is the difference between beer and ale?" As the guide acknowledged, while "the case involved the beer brewing business," the legal issues before the Court did not turn at all on the distinction between beer and ale. The Justice was simply curious.[5] Milkey took this all to heart: in a case like *Massachusetts v. EPA,* the Justices would expect him to both know and be able to explain climate science.

But the most important questions he would need to be prepared to answer were the notorious "hypothetical" questions, when a Justice tests the counsel's legal argument by positing facts—sometimes exceedingly farfetched—in order to explore the argument's limits. The hypothetical's purpose is to determine whether the advocate can demonstrate that the legal ruling that his side is asking the Court to deliver will not only lead to a legally defensible, just, and fair result in the specific case before the Court, but to legally defensible, just, and fair results in all future cases. During the oral argument in 2012 concerning the constitutionality of the Affordable Care Act's mandate that individuals purchase health insurance, Justice Antonin Scalia famously asked the solicitor general whether the government's legal argument in defense of the mandate's constitutionality meant that the federal government could also lawfully require people to purchase and eat broccoli.[6]

The Justices are well aware that in deciding cases they are establishing binding precedent that will govern rulings by other courts in hundreds, if not thousands, of future cases. So they naturally want assurances that their ruling will not cause unanticipated future problems. That is why an attorney who makes the classic mistake of responding to a Justice's hypothetical with the weak answer, "But those aren't the facts in this case!" will be invariably and swiftly rebuked: "I know that's not this case. Now answer my question!"[7]

The third task that Supreme Court advocacy experts advised Milkey to undertake was to develop two or three "themes." They explained that the Justices would be meeting in a private conference to deliberate and decide the case within a day or two of the argument. Milkey had to determine the few take-aways or sound bites from the case he wanted them to be thinking about as they walked off the bench and into deliberations. He needed to develop themes that would maximize the possibility of a favorable vote by a majority of the Justices.

Every case offers multiple possible themes. Which was the best for the *Massachusetts* petitioners? Was it the compelling nature of climate change? Was it the need to safeguard the ability of states to protect the interests of their citizens? Was it protecting the will of Congress from being flouted by an agency official? Milkey needed to decide which theme had the best chance of capturing a majority. Once he had picked his themes, Milkey's challenge was to craft ways to work them into his argument.

Because of the frayed relationships within the petitioners' core legal team, Milkey did not meaningfully consult with his several co-counsel in preparing for oral argument—not even Heinzerling or Mendelson, who would sit next to him at counsel table during the argument. He and Heinzerling were no longer on speaking terms in the wake of the controversy over who would present the oral argument. Their decade-plus-long friendship had ended. Instead, he relied almost exclusively on the judgment, wisdom, and personal loyalty of Carol Iancu and Bill Pardee, two exceedingly talented attorneys within the Massachusetts Attorney General's Office with whom he had worked closely throughout the litigation. Doniger complained openly about his lack of input into and knowledge about Milkey's progress.[8] Bookbinder shared Doniger's frustration at being frozen out. But they had no effective recourse—and Milkey felt no obligation to follow or even acknowledge the advice they were sending him by email.

The Binder

Milkey relied heavily on a good old-fashioned loose-leaf binder to collect all of his work product in support of his preparation for oral argument. It wasn't much to look at: a standard three-ringed white plastic binder that a parent buys a middle-school child for a few dollars from the local drug store on a back-to-school special. Certainly nothing about the binder's outward appearance suggested it was headed to the advocate's lectern facing the nine Justices of the Supreme Court.

"Much blood sweat and tears," though, as Milkey put it, went into preparing the binder, which "was really the whole MO of prep." He constantly revised its contents in light of new research and his new thinking from early October until the morning of the argument itself in late November.[9] Milkey's final binder reflected the argument he had developed over time. It included the affirmative points he wanted to make, answers to the hardest and most important factual and legal questions he anticipated that the Justices might ask, and the themes he hoped would stick with the Justices as they walked down from the bench at the argument's end.

By Milkey's account, he went through about one hundred versions of the content of his binder. He boasted that he "wore out the printer" in his office in the process, printing new versions two or three times per day—and, like the president's nuclear football, "the binder never left his side."[10] The final version had nineteen tabs and consisted of thirty-seven pages, printed double-sided so Milkey could see two pages at a time when he looked down during a practice argument. The first seven pages were all on the standing issue. Only after those seven pages did the binder include a one-page introduction, followed by three pages on why it was "common sense" that greenhouse gases were air pollutants.[11]

The binder's typewritten text was printed in four different colors of ink. The main points were in black ink and in a larger font. The subsidiary points appeared in red, green, and blue, with a new color for each new point. The wording was abbreviated to fit more text in, and in the margins, he jotted down a few handwritten additions in red ink. The binder's one-page argument introduction provided:

—M-C-J & M-I-P-T-C [black ink]
—If I may, I'd like to frame the merits v quickly & then turn immed. to stdg [black ink]
—Although the case before you arises in an imp. policy area, [blue ink]

—it turns on ordinary principles of stat interp. & admin law
[red ink]
—EPA made a decision based on two grounds, both of which=
plain errors of law [green ink][12]

The largest number of binder pages set forth answers to the questions Milkey anticipated from the Justices. His answers were typically no longer than three lines, to allow him to deliver them quickly, each line in a different color of ink. "Isn't this a global problem better addressed internationally?" one question posed. And "Doesn't the word 'judgment' in the Clean Air Act provide the EPA with broad discretion on when to make an endangerment determination?" And would "bullets" or "oxygen from plants" be air pollutants in your interpretation of the meaning of air pollutants under the Clean Air Act?[13]

After carrying his binder with him everywhere he went, on argument day Milkey brought it with him to the Court, and placed it on the lectern. Once the argument began, he never looked down at it once. The best advocates almost never do. After that first sentence, the binder's work is done.

"Each Moot Had Its Own Level of Pain"

As with any performance, rehearsals for Supreme Court oral arguments are essential. They take the form of a "moot court" where the advocate presents argument before a panel of experienced advocates who play the role of Supreme Court Justices. The "Justices" are veteran Supreme Court advocates and lawyers who are expert in the specific legal issues before the Court. An outstanding practice argument should be even tougher than the real thing.

The moot court consist of two parts and lasts a few hours. During the first half, the advocate and the "Justices" stay in role. The Justices ask tough questions, and the counsel does her best to answer.

During the second half, all the participants engage in an informal conversation about which arguments were persuasive or not, and what the advocate needs to do to improve.

The conversations are candid, and sometimes brutally honest. The mock Justices pull no punches in making it clear where they think the advocate's arguments and answers fall short. And if they believe the advocate will lose the case, they will tell them so and discuss strategies for making that loss the least harmful one possible: what experienced Supreme Court advocates have dubbed "a soft landing."

Milkey organized three moot courts. The first was on October 31, and was arranged by the National Association of Attorneys General in D.C. The second was on November 10 in Boston, co-sponsored by Harvard Law School and Suffolk University Law School. And the third moot was on November 17, at Georgetown University's Supreme Court Institute back in the nation's capital. The Georgetown moot was deliberately scheduled as the last because of its deserved reputation as the toughest.

Milkey described each moot as having "its own level of pain."[14] He entered each one "thinking he was fully prepared only to find out each time that he was not."[15] At the first moot, he began to appreciate for the first time that the petitioners had a major standing problem. All the "Justices" made it clear that the solicitor general's brief had persuaded them that the petitioners were at risk of losing on standing. It was sobering news, given that a loss on standing would be the most damaging way to lose the case; it could deny the ability of any climate plaintiff to bring a lawsuit in federal court in the future. In other words, not only would the *Massachusetts* petitioners lose their case, but all future climate plaintiffs would be cut out of the federal courts.

The second moot was even more demanding. The Boston "Justices" pummeled Milkey with questions. They included Jody Freeman—a Harvard Law School professor expert in environmental law—and also private practitioners who had either clerked for or argued before the Supreme Court. The Justices were not shy about making clear where they thought Milkey's argument needed more work or to be fundamentally rethought.

Those in attendance had widely different reactions to the Boston moot. Milkey came away feeling battered but not broken. He and his colleagues from the Massachusetts Attorney General's Office thought his answers still needed work, but that there was time to improve their content and his delivery. The argument was still two and a half weeks away. But other Carbon Dioxide Warriors who had long opposed Milkey's presenting the argument were more concerned than ever. Some had traveled to Boston to attend the moot in person and had passed judgment from their seats in the jury box. Others watched the moot live-streamed on the internet or listened in on conference call lines.

Doniger commented afterward to one of the attorneys who had served as a "Justice" that each of Milkey's moots "was worse than the one before."[16] He further suggested that they needed to "figure out how to do [a] coup"—meaning knock Milkey out of the oral argument.[17]

Bookbinder reported to his colleagues that another one of the moot court "Justices" repeated to him the old adage that you did not win cases with an excellent oral argument, but sure as hell you can lose a case with a poor argument. According to Bookbinder, that same attorney told him Milkey's argument is "costing you votes."[18]

Five days later, Milkey had his last practice argument: a full dress rehearsal for his argument at Georgetown. The Georgetown Supreme Court Institute, which is widely celebrated for its moot court program, provides rigorous practice arguments for attorneys in almost every case that goes before the Court. The institute boasts a room that replicates the exact dimensions of the actual courtroom at the Supreme Court in terms of the positioning of the lectern and the bench. Even the carpet on the floor and the clock behind the Justices' bench are virtually identical in appearance to those in the Supreme Court.[19]

A signature strength of the Georgetown moot court is that the attorneys serving as the Justices are not cheerleaders for the side they are questioning. Just the opposite. Georgetown places a premium on bringing in Justices who are skeptics of the position, so that they will be far more likely to anticipate the questions of a hostile Justice. The Georgetown Justices are known for being demanding and unforgiving in their follow-up questions. The Supreme Court Institute's moot courts are also strictly confidential, allowing both the Justices and the arguing counsel to be candid in their discussions afterward about the weaknesses in the case and possible grounds for improvement.

The Georgetown "Justices" included several members of its law faculty, one of whom had an unusually deep familiarity with the case: Lisa Heinzerling. The intense awkwardness that existed between Heinzerling and Milkey had to be put aside, given that the moot court was taking place where she was a highly regarded member of the law faculty as well as an obvious expert on the case. She could not be denied a seat on the moot court bench. But there was no remaining pretense of personal friendship or professional collegiality. By that point Milkey had cut off all contact with Heinzerling, as well as with Doniger and Bookbinder, while all three had remained in close contact with each other, sharing their rising alarm about Milkey's abilities. The moot court was essentially Heinzerling's only opportunity to provide Milkey with advice to which she knew he would listen.

The other four Justices were a virtual who's who of expert Supreme Court lawyers who had no problem playing the role of skeptical Justices. They included Louis Cohen and Tom Merrill, two former deputy solicitors general of the United States, and Bartow Farr and Richard Taranto, two former assistants to the solicitor general. All had argued many times before the Court and had clerked for Supreme Court Justices, and most were politically conservative.

All the Justices were unrelenting, firing off more than 120 questions in one hour. The first 23 questions were all about standing. But they also asked tough questions about whether greenhouse gases were air pollutants. The Georgetown moot Justices insisted on knowing whether Milkey's legal theory meant that oxygen could be considered an air pollutant. Further, they probed the logical outcome of the greenhouse gas argument. If carbon dioxide was an air pollutant, as Milkey contended, did that mean that people, who after all emit carbon dioxide when they breathe, were "mobile sources" of pollution subject to Clean Air Act regulation?[20]

Following the moot, the Justices offered Milkey advice on how to improve his odds. They urged him to focus on injuries to Massachusetts to establish standing, and to stress the state's loss of coastal territory from climate change. They further advised Milkey that his best, if not only, way to win on the tough question of whether the EPA had the discretion to decide to postpone the endangerment determination was to make the very narrow argument that the EPA's mistake was its reliance on an illegitimate reason—its policy disagreement with Congress concerning whether greenhouse gases should be regulated—rather than on any of the many lawful reasons the agency might have offered instead.

They stressed that his chances of prevailing would go up dramatically if he made clear the limited nature of the relief he was seeking from the Court, and they strongly suggested that he emphasize that the petitioners were not asking the Court to order the EPA to regulate greenhouse gases. They were merely asking the Court to compel the agency to reconsider its prior decision to postpone the endangerment determination, and to exercise its judgment based on "proper factors."[21]

Right after the moot, Milkey was dead tired, but despite the bruising exchanges, he remained positive. The moots had been hard, but also immensely helpful. For the first time in weeks, he was beginning to feel more confident about the argument.

"For the Good of the Case"

Milkey had not spent significant time after the moot debriefing with Bookbinder, Doniger, or any of the other Carbon Dioxide Warriors who had attended the moot court as observers. Under Georgetown's rules, members of the petitioners' legal team were not permitted to ask any questions or participate in the comment session. But that does not mean they didn't have a reaction. Doniger and Bookbinder were now more worried than ever. They thought the Georgetown moot had gone poorly and were increasingly frustrated by their lack of access to Milkey. They had sent him emails offering specific advice, including questions to anticipate and possible effective answers. But they had no confidence that Milkey was paying attention to their concerns or advice.[22]

This was also their first experience with the rigor of a moot court like Georgetown's. As a result, they did not fully appreciate that even expert advocates with many Supreme Court arguments under their belts often struggle mightily. It is very rare to see an advocate, even the most skilled, handle the barrage of questions without struggling.

That is, after all, the purpose of a high-quality moot. The attorney subject to a challenging moot learns from past mistakes. Bookbinder and Doniger, however, lacked the underlying trust in Milkey required to appreciate the potential for that learning process. Because they were walled off from contact with Milkey, their worries only deepened, given the importance for climate change of the *Massachusetts* case.

Doniger and Bookbinder brought their concerns to other members of their small group. They all reportedly agreed it would be best if Milkey was replaced.[23] They wanted Heinzerling to do the argument, though Bookbinder—who had never argued before the Supreme

Court—privately speculated that he might have to step in and argue the case. The Massachusetts attorney general had already decided against Heinzerling's presenting the argument and, as Bookbinder described it, Doniger, even though "he's a great thinker," "does not play well" in court because of his "refus[al] to relent on anything and a certain amount of arrogance."[24]

It was Bookbinder's responsibility to call Milkey. He reached him at home two days after the Georgetown moot to inform him of their decision: "Jim, I want you to step aside." "For the good of the case."[25]

Bookbinder later described the call as one of the hardest things he has ever done.[26] He and Milkey had been close friends for many years and he knew that Milkey would be angry, possibly furious. But "for the good of the case," which was far more important than the needs of any one person, Bookbinder felt he had no choice but to make the call.

Bookbinder guessed correctly. Milkey felt betrayed by Bookbinder, later equating Bookbinder's request with being stabbed in the back. Doniger, he thought, at least could be counted on to "stab you from the front."[27]

Milkey flatly and angrily rejected Bookbinder's request, but he could not help but be momentarily destabilized. It was not remotely plausible that a new advocate could come in at this late date and be able to do all the work necessary to be prepared for a Supreme Court oral argument—an enormous challenge that requires weeks and hundreds of hours of intense preparation and practice.

Yet the preposterousness to Milkey of Bookbinder's request also underscored the depth of Milkey's problem: Bookbinder—and apparently others on the team—clearly harbored a deep lack of confidence in Milkey's abilities. Like Milkey, they cared deeply about the case. Doniger had worked on the climate issue for more than two decades. They simply did not believe Milkey was up to the challenge.

The Georgetown moot, however, was not Milkey's last practice argument in front of an audience. On Tuesday, November 28, the day before the actual argument, Milkey did one last moot. This time, he stood before a very different kind of "judicial bench."

Unlike Georgetown's, this bench had no pretense of resembling the one in the Supreme Court's actual courtroom. It was instead outdoors, one of the dozens of identical gray-green wooden benches that have lined the pathways along the National Mall's vast lawn since the 1930s. Somewhere on the pathway between the Washington Monument and the U.S. Capitol, which is itself immediately across the street from the Supreme Court, Milkey presented his final moot. It was a clear, warm day for late November, with temperatures reaching the low sixties.

There was only one "Justice" sitting on the bench that afternoon. The Justice for the day, Milkey's wife, Cathie, asked no questions during his argument. When he was done, she simply said, "You know this."[28]

14

Seventy-Four Inches

On November 29, Jim Milkey and his opposing counsel, Deputy Solicitor General Gregory Garre, woke early. Milkey was staying at a Holiday Inn less than two blocks from a homeless shelter not too far from the Capitol—the Commonwealth of Massachusetts has a limited daily allowance for reimbursable travel expenses. Garre was at his home in northern Virginia, where he lived with his wife and daughter, who had celebrated her first birthday only a few weeks before. Both men followed the same time-tested ritual for "argument day." They took especially long hot showers. And in the shower, each one practiced his oral argument out loud one last time.[1]

Moses Humes Jr., the assistant supervisor of the Marshal's Aides, arrived at the Court well before either one of them. Humes had joined the Marshal's Office in 1999, after a long career in the military. He first visited the Court in 1993, when Justice Thurgood Marshall had been lying in repose in the Great Hall, never dreaming he would one day work there. But after several decades of service in the Air Force and Army—as an administrative assistant from 1956 to 1960 and then as a medical specialist from 1977 to 1999—he had returned to the Court at sixty to report for his new line of duty.[2]

Humes was still military fit, and he lit up rooms with his brilliant smile and vibrant ring of white hair. He was beloved at the Court. On argument days, it was Humes's job to ensure that the Supreme Court courtroom was ready by 10 a.m., when the arguments would begin.

Those employed at the Court take enormous pride in their jobs.[3] The image the Court strives to portray to the outside world—that of an institution that must be scrupulously fair, careful, and highly professional—is an essential ingredient in the success of a legal institution that possesses authority only so long as the other two branches of government and the general public accept its judgment. With no formal power to enforce its rulings, the Court must cultivate power through its prestige.

The only Chief Justice to have served as president of the United States, William Howard Taft, understood this basic truth. That is why Chief Justice Taft insisted that the Court move out of the Capitol Building, where it had occupied space lent to it by the Senate (at one point the Justices were relegated to the Senate chamber basement) and where the Justices had sat since the early nineteenth century. How could a coequal branch of government be taken seriously if it was conducting its work in another branch's building and if, when space got tight, the Justices were literally under the feet of the members of that other branch? So, wielding the persuasive powers of a former president, Taft in 1929 (just before the Great Depression) secured the funding from Congress necessary to construct a truly magnificent building for the Supreme Court. Both its distinct location and its powerful design projected the authority of an independent judiciary. The contrast between the Court's old quarters and its new home, completed in 1935, five years after Taft's death in 1930,[4] was so great that several Justices initially balked at moving in. They said they would feel like "nine black beetles" moving into the ancient Egyptian Temple of Karnak and mockingly asked whether they would need to ride elephants to work every day. But there was a serious side to their jokes, because Taft's makeover of the Court building, initially championed by the architect Cass Gilbert, was designed in part to make everyone, including the Justices themselves, appreciate the importance of the Court's work.[5]

By the morning of the *Massachusetts* argument seventy years later, the Justices had become accustomed to the grandeur of their quarters. They looked the part in their deliberately austere black robes, designed to underscore their objective neutrality. It is an image that, to their dismay, contrasts sharply with how members of the public have increasingly begun to perceive them based on their voting record in many of the Court's highest-profile cases—as political partisans expressing the policy preferences of the president who nominated them. Ensuring that the Court presents itself to the public in a manner consistent with its self-image of rigorous neutrality was a responsibility that Humes and the Marshal's Aides he supervised took seriously. Every detail has to underscore the rigor and exactitude of the Court's work, its probity and lack of bias.

All nine black leather chairs—each with a Justice's name engraved on it—must be in their proper places behind the bench where the Chief Justice and the eight Associate Justices sit, positioned equidistant from each other and with the back of each chair set at the same angle and height. By each Justice's side is a green porcelain spittoon, which (happily) is no longer used for its original purpose and functions instead as a wastepaper basket.[6]

One by one, as he had long been accustomed to doing, Humes arranged all nine black chairs and green spittoons. Then, assisted by more junior Marshal's Aides, Humes opened a small wooden dresser behind the bench, carefully removed a tray holding nine pewter mugs, and placed each mug at its proper location.[7] The name of the Justice is engraved on the mug followed by the name of the Justice they replaced, going back in time, reminding each Justice of his or her own distinct lineage. Because pewter is a soft metal, the older names can be hard to decipher, which discourages the mugs' frequent washing. In their own simple way, the mugs remind the Justices of the historic but ultimately temporary role they each serve as a member of the Court; their name will not be the last name on the mug, and it will disappear over time.

Finally, Humes turned his attention to the two tables where the advocates sit before they rise to the lectern and present their oral arguments. Pads of paper, water pitchers, and water glasses must be available and properly positioned on each table. Four handcrafted writing quills are also laid out, precisely positioned, and configured. A small piece of paper, no bigger than an index card, offers a seating chart, indicating where each Justice, by name, sits on the bench.[8]

For the past seven years, Humes had repeated this ritual before every argument day—now about forty a year—as Marshal's Aides had done for more than a century, since the appointment of the first marshal in 1867 charged with providing security for the building and the Justices, safeguarding all Court property, and attending all of the Court sessions. When Jim Milkey and Greg Garre reached the courtroom several hours later, everything had been set out perfectly. Humes's job was done.

"Make Me Look Like the Biggest Asshole in the World"

As Milkey passed by the Court's famous façade and entered the Maryland Avenue side entrance, the day's unseasonable warmth—temperatures would reach sixty-two degrees—seemed a good omen, a reminder, even if wholly unscientific, that global warming and climate change were real. Milkey made his way through the first set of security scanners by the entrance and idled for a bit next to a larger-than-life-size bronze statue of Chief Justice John Marshall, while waiting for a member of the Marshal's Office to escort him upstairs. A self-described "cheapskate," Milkey proudly wore a suit he had bought at Costco, which he also declared to be a "good luck charm." For good measure, he had brought two further good luck charms— two stones, one black and the other pink—lodged in the suit's pocket. The stones had been given to him by one of his co-counsel from the Massachusetts Attorney General's Office; and although Milkey swore that he didn't believe in that kind of thing, he thought, "Why not take a chance?"[9]

The Sierra Club's David Bookbinder, champion of the failed effort to block Milkey from arguing the case that morning, approached him by the Marshall statue. No doubt seeking to inspire, Bookbinder came up with a curiously backhanded compliment. "Jim," he said, "your job today is to make me look like the biggest asshole in the world."[10]

Milkey did not respond. He didn't even blink. He was as motionless and dispassionate as the statue he stood next to. He was, he later said, completely "in the zone."[11]

Nor did he react to or even later recall the woman in her early fifties, in a dark suit, who quietly approached him, shook his hand, and stated simply, "May the best arguments win." Unbeknownst to Milkey, she was one of the many career attorneys who had worked on the EPA opinion denying Mendelson's petition several years before. And like many EPA career attorneys, she was privately rooting for Massachusetts to win on the question of whether greenhouse gases were air pollutants.[12]

Her hopes might have dimmed on close appraisal of Milkey. He had slept so little in the two weeks before that his wife, Cathie, told him he "looked gray." She was referring to the pallor of his skin, not the color of his hair. Neither Cathie nor their two young teenage sons felt like they had seen him for weeks, if not months, even when he was nominally at home. The case had consumed all of his time and energy, leaving no room for anything else.[13]

A few minutes later, escorted by the deputy clerk of the Court, Milkey entered the courtroom after first briefly meeting with William Suter, the clerk of the Court, to go over morning logistics. Milkey had been in the courtroom before but never to deliver an oral argument. The room itself is massive: 83 by 91 feet with 44-foot-tall ceilings, its perimeter marked by twenty-four columns of Siena marble. The walls are of ivory vein marble from Spain, and the floor borders are Italian and African marble. The courtroom had been built at the heart of the Great Depression, but ironically that had been a boon for the Court because Congress had authorized $9,740,000 for the new building's construction just before the 1929 stock market crash, which had significantly lowered the actual construction cost. As a result, there was ample money available to construct a building of exceptional beauty—and, when completed, the Court was still under budget and returned money to the U.S. Treasury.[14]

High above the bench where the Justices sit, a newcomer could spot a series of four 40-foot-long Spanish marble friezes—reportedly from a quarry near Monovar in the Province of Alicante[15]—each one underscoring the sanctity of the rule of law. Directly above the bench on the east wall are two glowering bare-chested figures in Roman togas representing the "Majesty of Law" and the "Power of Government." They sit side by side, separated by a tablet with Roman numerals I through X, representing the Constitution's Bill of Rights. To their left and right are figures representing "The Defense of Human Rights and the Protection of Innocence," and the "Safeguard of the Liberties and Rights of the People in their Pursuit of Happiness." On the frieze immediately behind them, an American bald eagle defiantly spreads its wings.[16]

For lawyers, a Supreme Court argument is an advocate's grandest stage—an opportunity to join the historic ranks of Daniel Webster, Henry Clay, and Thurgood Marshall who once stood, like them, at the lectern before the Court.[17] But any lawyer about to present oral argument before the Justices who mistakes himself for the equivalent of the Olympic athlete before a contest quickly discovers that the more apt analogy is that of an untrained *bestari* fighting wild animals in the Colosseum in Rome.

The Justices' barrage of demanding, wide-ranging, and penetrating questions is a clear reflection of their high expectations of the counsel who appear before them. For anyone unprepared or unable to answer the Justices' questions, there is no place to hide. Any weaknesses in their arguments will quickly be exposed and their proffered reasoning shredded. The unrelenting intensity of questioning has caused some advocates to lose their breath or even faint dead away. In trying to defend President Franklin Delano Roosevelt's New Deal legislation, U.S. Solicitor General Stanley Reed passed out under a torrent of questions unleashed by the Justices, led by Felix Frankfurter. An attorney once fainted in the exact same spot where his father had fainted decades earlier.[18]

When Milkey was ushered in, the courtroom was quickly filling up to capacity. Everyone knew that *Massachusetts* was a big case, by some accounts the most significant environmental case ever heard by the Supreme Court. Members of the public and of interest groups on all sides of the climate issue—even attorneys who had worked on the case—had camped out overnight on the sidewalk in front of the Court to secure one of the 250 coveted seats. Anyone who made the mistake of arriving after 6 a.m. was far too late. At best they would be relegated to the "three-minute line" at the very back of the courtroom, where a dozen or so people were allowed to watch the proceedings for approximately three minutes before being escorted out and replaced by the next group.

Also quickly filling up to capacity were the eighty seats for members of the Supreme Court Bar, at the front of the courtroom immediately behind the counsel's table, separated from the public section by a bronze railing. Some of the industry lawyers who had come that morning could bill $500 or more per hour for the time they spent in line (beginning at 6 a.m.) to attend the argument. Anyone showing up at the Bar line after 7 a.m. was too late for a courtroom seat, and would be shunted off to a room reserved exclusively for members of the Bar where they could listen to a live audio feed.

To the far left of the bench is the area reserved for the national news media, and it, too, was filled to capacity—seventy people can just about squeeze in, though anyone in the back will have an obstructed view. The seats are assigned by seniority, so veteran Supreme Court reporters like Linda Greenhouse of the *New York Times,* Robert Barnes of the *Washington Post,* and Nina Totenberg of *National Public Radio* had their seats of honor up front. They surveyed the scene, scanning the crowd for anyone who might be worth following up with later.

By 9:59 a.m., only nine seats remained empty. Unbeknownst to those nervously awaiting their appearance, the nine Justices were already in the courtroom, hidden from view by 26-foot-tall, dark red, heavy velvet curtains trimmed in gold immediately behind the bench. Several likely joked quietly about the unseasonably warm weather outside. Then, each Justice, in a long-standing tradition established in the late nineteenth century by Chief Justice Melville Fuller, shook the hand of the other eight. Fuller became Chief in 1888 at a time when the Court was reportedly populated by Justices who were highly individualistic and prideful, and the handshake was one of several ways that Fuller was credited with avoiding acrimony and rifts within the Court.[19]

At precisely 10 a.m.—marked by a final click of the minute hand on a large bronze clock designed, like the building itself, by Cass Gilbert and located directly above the bench—a small chime sounded in the already hushed courtroom. Marshal Pamela Talkin banged her gavel, and all nine Justices emerged simultaneously from three spots behind the curtains. Everyone in the courtroom rose to show their respect.

Harkening back to the Anglo-Norman word "to hear," the Marshal announced:

Oyez! Oyez! Oyez! All persons having business before the Honorable, the Supreme Court of the United States, are admonished to draw near and give their attention, for the Court is now sitting.

The Court was now in session.

The Nine

Now in his second year on the job, Chief Justice John G. Roberts Jr. took his seat in the middle of the bench. Only the seventeenth person to serve as Chief Justice of the United States, Roberts was a youthful fifty years old at the time of his appointment, the youngest Chief Justice since John Adams appointed forty-five-year-old John Marshall to the Court in 1801. Roberts had a four-year-old son and a five-year-old daughter—the youngest children of a Chief Justice since Oliver Ellsworth's appointment in 1796—prompting the Chief's security detail to purchase two car seats at a local warehouse toy store for the large black SUV they used to transport him from his home in suburban Maryland—a perk Chief Justice Ellsworth had not enjoyed.

Roberts had originally been nominated by President George W. Bush in July 2005 to take Justice Sandra Day O'Connor's seat on the bench after she announced her retirement earlier that month. But three days before Roberts's scheduled Senate confirmation hearing, Chief Justice William Rehnquist died—and President Bush nominated Roberts to the Chief Justice slot instead. Roberts had served as a judge on the D.C. Circuit only a couple of years at the time of his nomination, but he had impeccable professional and political credentials and seemed to be on his way to easy confirmation for O'Connor's slot when Bush made the switch because Roberts had so impressed Bush with his leadership potential.

Roberts had graduated at the top of his class at both Harvard College and Harvard Law School, had served as clerk on the U.S. Supreme Court, and had been widely considered to be one of the nation's most highly gifted Supreme Court advocates. He had been a political appointee in the White House and Justice Department during the administrations of Presidents Ronald Reagan and George H. W. Bush, but he did not have a reputation as being highly ideological or politically partisan. His strong performance during his confirmation hearings, in which he impressed the nation with his intelligence, charm, and wit, led to easy confirmation by a lopsided Senate majority in otherwise sharply divided political times. During those hearings, he promised the nation that as Chief Justice he would set a tone of judicial modesty, encouraging unanimity and less divisiveness in voting and more narrowly drawn rulings. At the time of *Massachusetts,* the Chief was sufficiently new on the bench that his promise had not yet been tested.

Seating is determined strictly by seniority. To the Chief Justice's immediate right that morning was Associate Justice John Paul Stevens, who had served on the Court for thirty-one years. To Roberts's immediate left sat Antonin Scalia, who had served on the Court for twenty years. Both had been nominated by Republican presidents—Stevens by President Gerald Ford and Scalia by President Reagan—and both had been confirmed by the Senate by identical votes of 98–0. But once on the beach, the two Justices voted very differently in the most high-profile cases, with Justice Stevens increasingly voting in ways characterized as liberal and Justice Scalia as a stalwart conservative. Their personalities on the bench could also hardly have been more different. While each was capable of unraveling an advocate's arguments, the midwesterner Stevens was invariably polite and succinct in his questions, routinely asking the advocate for permission to interrupt. Hailing from Queens, Scalia, by contrast, captivated the courtroom with his loud and demanding voice and seemingly relished the opportunity to engage in verbal battle with those arguing before him.

The remaining six Justices sat alternating to the Chief's right and left in decreasing order of seniority: Justices Anthony Kennedy,

David Souter, Clarence Thomas, Ruth Bader Ginsburg, Stephen Breyer, and Samuel Alito, who sat to the Chief's far left. Justices Kennedy, Thomas, Souter, and Alito had each been nominated by Republican presidents—Reagan, George H. W. Bush, and George W. Bush. Of the four, only Justices Thomas and Alito had proven to be as conservative on the bench as their billing when nominated. Thomas enraged liberals because he had replaced Thurgood Marshall and consistently voted in ways wholly antithetical to Marshall, who was a liberal icon. Alito was still fairly new on the Court, having joined the Court only a few months after Roberts, and his principal qualification was that his sixteen years of rulings on the federal appellate court in Philadelphia left no doubt that he would be a reliable, conservative vote. Thomas's only surprise on the bench had been his policy of almost never asking questions during oral argument, even though his opinion writing made it clear he was fully engaged in the cases and had strongly held views. As the most junior Justice, Alito was asking relatively few questions, but what was striking was the sometimes almost grouchy way he posed them. Unlike someone like Scalia, Alito rarely seemed to enjoy the arguments, and he lacked Stevens's politeness and polish.

Kennedy and Souter were the reason Alito's apparent reliability as a conservative vote had been so important to Republicans. When he was nominated by President George H. W. Bush in 1990, Souter was a complete unknown nationally, having served on the federal appeals court in Boston for less than five months after seven years on the New Hampshire Supreme Court. His lack of a paper trail had been considered an asset to secure Senate confirmation, as the Democrats were in the majority—the president's chief of staff reportedly privately advised worried Republicans that Souter would be a "home run" for placing a conservative on the bench. But once on the Court, Souter tended to vote so frequently like Justice William Brennan, the liberal Justice he had replaced, that "No More Souters" became a battle cry for Republicans.[20]

Although not as liberal as Souter, Kennedy also had disappointed conservatives. He had generally voted in ways favored by conservatives, but he also harbored grandiose views of the role of the Supreme Court in advancing justice, which prompted him occasionally to vote for positions they deplored. Siding with Souter and the most liberal Justices on the Court, Kennedy had voted against overruling the Court's abortion decision in *Roe v. Wade*,[21] in favor of striking down as unconstitutional a Texas law outlawing homosexual activity,[22] and against imposing the death penalty on offenders who were juveniles or had intellectual disabilities.[23]

Both Kennedy and Souter could be active questioners at oral argument in cases that interested them, but they had very different styles. Sounding very much like the constitutional law professor he had once been, Kennedy was more likely to pontificate about his own views than to pose a series of questions designed to make a point. Souter was the opposite. He would ask a series of a questions with a threshold politeness that could be misleading. If he concluded that the attorney's answers were evasive, Souter could quickly become tenacious and demanding.

President Bill Clinton had nominated Ginsburg and Breyer to the Court during his first two years in office. Both were easily confirmed, by votes of 96–3 and 89–6, because they were largely considered to be judicial moderates despite their obvious liberal leanings. Ginsburg, of course, had been a pathbreaking champion of women's rights as a legal advocate, including successfully arguing six cases before the Supreme Court. But while a judge on the D.C. Circuit, she enjoyed a reputation as a fair-minded stickler for legal rules who did not have an ideological or political agenda. She received strong support from the business community. The same was true for Breyer. Formerly a professor at Harvard Law School, he had previously worked for the Senate Judiciary Committee, where he had impressed Democratic and Republican leaders with his evenhandedness and dislike for excessive federal regulation. While a judge on the U.S.

Court of Appeals for the First Circuit, Breyer similarly had a moderate judicial track record. But once on the Court, Ginsburg and Breyer both had increasingly become champions, by their votes and in their written opinions, of outcomes favored by liberals, especially Ginsburg on matters touching on gender equality.

They could both be active on the bench in cases that interested them. Breyer was infamous for his long-winded questions and Ginsburg for both her quiet voice and her extraordinary command of legal procedure. But everyone also knew her voice remained quiet only in terms of its decibel level.

"Draw Near"

Before the argument, Milkey had of course studied all the backgrounds and predilections of the Justices, and what immediately struck him as they entered the courtroom was how physically close they were to where the lawyers sat and stood to argue. Closer than in any courtroom he had ever been in. It was surprising, given the room's grandeur and solemnity.

Seventy-four inches is all that would separate the lectern behind which Milkey would stand and the mahogany bench behind which Chief Justice Roberts sat, flanked by four Justices on either side.[24] As "admonished" by the marshal in her opening declaration, lawyers who are about to present oral argument before the Supreme Court do in fact "draw near."

The surprisingly intimate setting—the Justices and counsel are almost close enough to reach out and touch hands—might seem to invite an informal conversation. But Milkey was not about to forget the formality of the setting, or the enormity of the stakes. The bench's apparent friendly embrace could quickly become suffocating.

The advocate's proximity to the bench presents an unexpected challenge. The Justices are so close to the lectern that the advocate cannot see all nine of them at the same time. Arguing counsel must turn from side to side to see the entire bench. But because of the acoustics of the room, the lawyer hears the questions being posed by the Justices only from speakers high above the bench. As a result, the counsel cannot identify the speaker by relying on the kinds of visual or auditory cues that one subconsciously assumes will be available when talking to another person at such close proximity. First-time advocates, unprepared for this acoustical trick, can quickly become confused during argument as they hear a question but cannot immediately identify which Justice has asked it. Sometimes the Justice asking the question will wave a hand to signal to the lawyer or even say out loud: "I'm over here." But sometimes they don't.

Seasoned advocates, aware of the problem, prepare by memorizing the voices of the Justices ahead of time, training themselves to listen for Souter's aristocratic Yankee dialect with its muted "r's"; Ginsburg's quiet, classic Brooklyn accent; Scalia's unmistakable booming voice from Queens; or the Chief Justice's flat vowels, suggestive of his Indiana roots. More-experienced advocates also know to scan the bench for the subtle physical sign that a Justice is about to ask a question: each one must lean forward ever so slightly to push the button on the microphone in front of them.

As Milkey surveyed the bench in those few seconds between the time the Justices entered and the moment the argument would begin, Justice William J. Brennan Jr.'s famous "Rule of Five" hung in the air.

Brennan had served on the Court from 1956 to 1990, and according to the lore of his chambers he would gather his new law clerks in his office soon after their arrival and quiz them by asking, "What is the most important rule of constitutional law?" Determined to impress their new boss, the young clerks would compete over the relative importance of such fundamental constitutional guarantees as due process, equal protection, free speech, the right to vote, the privilege against self-incrimination, or the prohibition on cruel and unusual punishment.

Justice Brennan would wait for each one to express a view before calling them all up short and declaring it was the rule of five. Brennan would then explain to his inevitably perplexed young charges that the most important rule of constitutional law, at least for their purposes over the course of the year in his chambers, was that it took five votes to secure the majority needed for an opinion of the Court.[25]

Milkey knew that he needed five Justices to win his case—and he did not expect that securing that razor-thin majority would come easy. At least four Justices the previous June had voted to grant review, but Justices who vote in favor of granting review are saying by their vote only that the legal issues presented in the case are sufficiently important to warrant the Court's attention, nothing more. Moreover, because the identities of the Justices that vote to grant review are not publicly disclosed, Milkey could not safely assume that the Justices who had voted in favor of review in *Massachusetts* had done so because they thought his legal arguments had merit. The four could have included Justices who saw the case as an opportunity to rule *against* environmental petitioners' standing to bring this kind of lawsuit.

Milkey had to put those dark thoughts aside and focus on the task at hand: framing his argument and answering the questions as best he could to win the case. Justice Kennedy, seated to the immediate left of the Chief Justice, would likely cast the key vote. And getting Kennedy's vote, Milkey had concluded, would turn on issues utterly unrelated to greenhouse gases. His argument that carbon dioxide emissions were air pollutants, Milkey thought, might well have traction with a large number of Justices, but first he would have to overcome the jurisdictional hurdle of persuading a majority of the Justices that Massachusetts and the other petitioners possessed standing to bring the lawsuit forward in the first place. On that issue, Kennedy was clearly the wild card of the nine.

Notwithstanding the Court's effort to promote the image of nine neutral judges, Milkey knew, based on each Justice's track record of voting, that all nine votes were not equally in play. The competing legal arguments made by the opposing counsel were likely to be received differently by the more conservative Justices than they would be by the more liberal Justices. Those factors are relevant in far fewer cases before the Court than members of the public assume, but *Massachusetts* was a case where the conservative/liberal divide was more relevant than in most.

Chief Justice Roberts and Justices Scalia, Thomas, and Alito were almost certainly opposed to the petitioners' argument that they had the standing to bring their lawsuit against the EPA, and Milkey had reason to hope that Justices Stevens, Souter, Ginsburg, and Breyer would be more favorably inclined. The contrasting positions of those eight Justices were fairly predictable, based on both their past votes and the tendency of conservative Justices to apply standing requirements more strictly than liberal Justices. Kennedy was the ballgame—or, perhaps more accurately, Kennedy would determine whether Massachusetts was eligible to play the game at all.

In this respect, *Massachusetts* was similar to all other environmental cases that had been decided since Kennedy had joined the Court two decades earlier. In all but one of those cases, Justice Kennedy had sided with the majority. And in many of those instances, he had been the deciding fifth vote. At least in environmental law, but not only then, the way Kennedy went was the way of the Court. That is why some jokingly referred to the Supreme Court as the "Kennedy Court" rather than, as is traditionally done, by the name of the Chief Justice.

But much as he might like to, Milkey knew he could not focus all of his attention on Kennedy. He had to answer every question he was asked, however challenging or unfair its premise. A lawyer arguing a Supreme Court case does not have the luxury of ignoring a particular Justice whose vote he knows is a lost cause. If Milkey had the option, he would gleefully agree to give up Justice Scalia's vote in exchange for his silence during the oral argument. But to persuade

Justice Kennedy, he would have to answer successfully the combative questions of the Justice he anticipated would be the most hostile: Justice Scalia.

Yet what Milkey couldn't do with his words, he could accomplish to a certain extent with his eyes. An outstanding oral advocate does not communicate by words alone. Counsel is also simultaneously sending messages to the Justices through eye contact, or the lack thereof.

It's a trick every new teacher quickly learns in the classroom. As soon as the students anticipate that the teacher is about to ask a question, what happens? Everyone in the room immediately looks down at their desks, suddenly discovering something totally fascinating to examine on the desk. Why? Because it's a lot harder to call on someone who is not looking at you.

That same dynamic is at play in Supreme Court advocacy, and the best advocates exploit it to their full advantage. An advocate must answer any question any Justice asks, no matter how hostile or potentially disruptive. But in addition to spending more or less time answering any given question, a lawyer may look more or less directly at certain Justices. After responding to a question from Justice Scalia, Milkey had been instructed to pivot immediately to a less hostile corner of the bench. By shifting his gaze, Milkey was in effect inviting a new intervention, making it a tad harder for Scalia to ask a follow-up question and a tad easier for a friendlier Justice to jump in, or at least one, like Kennedy, whose vote was more likely in play.

At 10:02 the Chief Justice announced that the Court would "hear argument first today in 05-1120, Massachusetts versus Environmental Protection Agency." That was Milkey's cue. He rose from his seat and moved to the lectern less than one foot to his right, and said, as so many had before him, "Mr. Chief Justice, and may it please the Court."

15

A Hot Bench

As Jim Milkey began to speak, those in the Supreme Court Bar in the front of the courtroom shifted forward ever so slightly in their seats, eager to see which Justice would be the first to interrupt and ask a question. Most looked immediately to the one known for sporting a Cheshire-cat-like grin before ambushing his prey.

Justice Antonin Scalia had, after all, transformed oral argument. The number of questions the Justices asked almost doubled immediately after Scalia joined the Court in September 1986, increasing to an average of 104 questions per hour-long argument during Scalia's first year. By 2005, the year before *Massachusetts* was argued, the average number of questions asked by the Justices had ballooned to a whopping 156.[1]

From the minute he joined the Court, Scalia made his presence known. During the very first argument on his very first day on the bench, Scalia asked twenty-eight demanding questions.[2] And during one two-week period that Supreme Court Term, which runs October to October, he asked 30 percent of all the questions posed by the Justices. In case after case, Scalia peppered arguing counsel with questions, prompting Justice Lewis Powell, an old-school Virginian who was well known for his polite demeanor, to quip to Justice Thurgood Marshall: "Do you think he knows the rest of us are here?"[3]

As a former law professor at the Universities of Virginia and Chicago, known for his lively questioning of students in the classroom, aggressive interrogations came naturally to Scalia. But this was more than just the judicial equivalent of classroom banter designed to ensure a full airing of the legal issues presented in a case. It was strategic.

Scalia had discovered upon joining the Court that his opportunity to influence the other Justices was extremely limited. By tradition, the Justices do not discuss with each other a case before oral argument. As described by Justice Kennedy, "Before the case is heard, we have an unwritten rule: We don't talk about it with each other."[4] Even at the conference held one to two days after the argument, where they formally voted, there was no meaningful discussion. One at a time, each Justice would speak without interruption and announce his or her vote, along with a summary explanation. To his disappointment, the conference was "not really an exercise in persuading each other."[5] Worse still, because Scalia was the most junior Justice, he was the last to speak—before he had uttered a word, all the other Justices had already announced their vote.[6]

So Scalia pivoted to the oral argument, the first—and by far the longest—time the Justices ever spoke about a case in one another's company. As explained by Chief Justice Roberts, "We come to it cold as far as knowing what everybody thinks, and so through the questioning, we are learning for the first time . . . how [the other Justices] view the case and that can alter how you view it right on the spot."[7] To Scalia, the lesson was obvious. If he could use those sixty minutes wisely, he could press his views on the case to his colleagues *before* the conference vote. He quickly became the master of using oral argument to highlight the weaknesses in the position he disfavored, and the strengths of the one he supported.

There was nothing subtle about it. Scalia asked far more questions and far tougher questions of the attorney arguing the side he anticipated voting against than the side he expected to favor. His barbed, sometimes mocking questions highlighted weaknesses in a lawyer's argument. His loaded hypotheticals frequently trapped arguing counsel into answering in a way that made their legal position seem ridiculous or otherwise tied them in knots. As one of his former law clerks put it, stealing a line from one of the Justice's most famous dissenting opinions, "This wolf came as a wolf."[8]

When the attorney for the side Scalia favored rose to speak, the Justice would either be completely silent or ask questions designed only to reinforce the points the counsel was making. And if that attorney was struggling in his response to another Justice's questions or answering in a way that Scalia considered problematic, Scalia would jump in to rescue the attorney, cutting off the bad answer by saying something like: "I would have thought your answer to that question would instead have been . . . ," followed by "Isn't that right?" To which counsel, spotting the obvious lifeline, would immediately reply: "That is absolutely right, Justice Scalia."[9]

Milkey knew Justice Scalia would not be his friend on the morning of November 29. What he would toss Milkey's way would not be lifelines—more like hand grenades.

"Ordinary Principles"

After the opening obligatory nine-word nod to the Chief and the Associate Justices—"Mr. Chief Justice, and may it please the Court"—Milkey began his argument with two short and straightforward sentences:

> If I may, I'd like to frame the merits very quickly and then turn immediately to standing. Although the case before you arises

in an important policy area, it turns on ordinary principles of statutory interpretation and administrative law.[10]

Deceptively simple. All thirty-nine words in those two sentences were the product of months of intense and often heated debate regarding precisely how to frame the case for the Justices, after repeated failed moot court practice arguments. Although Milkey never once glanced down at his binder when he spoke, every one of those words appeared verbatim, sometimes abbreviated, in that binder under the "Intro" tab in black, blue, red, and green ink.[11]

The words were designed to maximize the odds of securing the five-Justice majority necessary to reverse the lower court ruling and remand the case back to the EPA to reconsider its decision not to regulate greenhouse gas emissions from new cars and trucks. And they were designed to minimize the risks of his argument being immediately derailed by the Justice whom Milkey anticipated would be most vehemently opposed to his position. The script that he had memorized was as planned as the opening lines of any Broadway production, even though its wording was deliberately designed, like any good theater, to sound as if it was being spoken extemporaneously.

Milkey knew that Justice Scalia would be all over him on the standing issue. Even before joining the Supreme Court, Scalia as a D.C. Circuit judge had publicly vented his frustration with the willingness of courts to allow environmental plaintiffs to bring the very kind of lawsuit Milkey was asking the Court to endorse in *Massachusetts*. Relishing in dishing out rhetorical punches, Scalia had declared that while such lawsuits, designed to ensure "strict enforcement of the environmental laws . . . met with approval in the classrooms of Cambridge and New Haven," they were not equally applauded "in the factories of Detroit and the mines of West Virginia." Less enforcement of environmental regulation, Scalia chided, would be "a good thing, too."[12]

Milkey was concerned that Scalia would consume most or even all of his thirty precious minutes with questions about standing. He could effectively filibuster Milkey's argument and never give him a chance to address the merits of the case—whether greenhouse gases were air pollutants—or otherwise address the concerns of other Justices, especially Kennedy. By beginning the first sentence with a request for permission—"If I may"—Milkey sought to reassure Scalia that he would get right to the standing issue, and only wanted a few seconds to frame the case on the merits for the entire Court.

The gambit worked. Scalia paused. He allowed Milkey to get to sentence number two and frame his argument. A small victory for Supreme Court advocacy.

The second sentence, which spoke to the merits of the case, was significant for what it did *not* say. There was no mention of climate change. No trumpeting of the extraordinary importance of the climate issue. No hint of climate change's compelling nature and the tremendous stakes for humankind and therefore why it was critically important for the EPA to exercise its existing authority under the Clean Air Act to address the issue.

Milkey included only an indirect allusion to the climate issue— "Although the case before you arises in an important policy area"— and then quickly made it clear that the "policy issue" was beside the point. All the Court needed to do to decide the case was to apply "ordinary principles of statutory interpretation and administrative law." In short, the extraordinary nature of the case, which petitioners had stressed to persuade the Court to grant review and which had prompted members of the public to camp outside the Court overnight, had now devolved into nothing more than a run-of-the-mill application of "ordinary" legal principles. No great legal precedent needed to be made to rule in Massachusetts's favor.

Just like the first sentence, the second sentence was the product of carefully plotted strategy. Milkey and the entire *Massachusetts* petitioners' team shared the judgment that the odds of winning the case went up the less he made the case about environmentalism and climate change and the more he made it clear that he was not asking the Court to establish groundbreaking legal precedent. The key vote here was Kennedy, a legal conservative who had always been a skeptic of aggressive federal environmental regulation.

But Kennedy also cared deeply about states' rights. This provided a useful theme for framing Milkey's argument in a way that Kennedy might find attractive: the idea that the federal government, notwithstanding clear statutory language, was failing to address an important policy concern of the states. Milkey accordingly sought to make the case less about environmental protection and climate change *per se* and more about the failure of the federal government to adhere to "ordinary principles" of law to address state needs.

Milkey closed his opening comments on a similar theme. He stressed the limited nature of the ruling Massachusetts was asking the Court to make:

> We are not asking the Court to pass judgment on the science of climate change or to order EPA to set emission standards. We simply want EPA to visit the rulemaking petition based upon permissible considerations.[13]

An environmental activist listening to Milkey's statement might have been shocked, if not appalled. Milkey's words made it seem as if he was deliberately tanking the case. He appeared to be unilaterally abandoning what many environmentalists would have naturally assumed the case was all about: to make clear that greenhouse gas emissions endanger public health and welfare and that the EPA was therefore required to restrict greenhouse gas emissions.

But Milkey knew what few environmental activists would appreciate: those were losing *legal* arguments. The Court would never make a scientific finding on its own that greenhouse gases endanger public health and welfare. Nor would the Court in this case ever order the EPA to find that greenhouse gases endanger public health and welfare and to restrict greenhouse gas emissions. Most certainly Justice Kennedy wouldn't.

What environmentalists would want to hear Milkey say was wholly irrelevant. The best environmental lawyers are not the best "environmentalists"—they are the best "lawyers." Milkey's job was not to present an argument that environmentalists wanted to hear and applaud. Those kinds of speeches could be saved for congressional hearings, political rallies, and campaign fundraisers.

His exclusive audience that morning was the nine Justices. They alone would decide the outcome of the case, and the kinds of political arguments climate activists might like to hear would fall entirely on deaf ears within the Supreme Court.

Milkey was a lawyer talking to lawyers about legal issues. He needed to make a successful legal argument crafted for the Court he actually faced, not the Court he might have preferred. The window for accomplishing that result was extremely narrow, and his best bet was to ask for less rather than more.

That meant making it absolutely clear that Massachusetts was *not* asking the Court to rule on the science of climate change. It further meant making it clear that Massachusetts was also *not* asking the Court to order the EPA to regulate greenhouse gas emissions.

A narrow argument admitted of the possibility that the EPA could, following such a Supreme Court ruling, go back and decide again on different, wholly legitimate grounds not to regulate greenhouse gas emissions. But that was a risk Milkey and his fellow Carbon Dioxide Warriors had concluded that he had no choice but to take. So Milkey told the Justices from the outset that all Massachusetts wanted was for the EPA to revisit Mendelson's petition "based upon permissible considerations." Milkey was deliberately downplaying his "ask."

A Supreme Court win in this case, on any basis, might later be transformed into an enormous political victory, just as a loss could be politically destructive. There is a time to talk big and a time to talk small. This was a time for the latter.

Having completed his framing of his legal argument and having stressed its limited nature, Milkey then told the Court: "And now, Your Honor, I'd like to turn to standing."[14]

"When Is the Predicted Cataclysm?"

Justice Scalia predictably pounced just seconds into Milkey's argument that the *Massachusetts* petitioners possessed the constitutionally required standing to bring its climate lawsuit in federal court. He asked nine questions on the standing issue in rapid succession, before any other Justice had a chance to chime in. And Scalia never relented. Every answer Milkey attempted to give was quickly interrupted by another question in the barrage Scalia unleashed: of the fifty-eight questions the Justices asked Milkey, Scalia asked twenty-three. His object was clear: to eviscerate Milkey's claim that the *Massachusetts* petitioners had the right to file a suit in federal court for relief from the consequences of climate change.

Scalia began by asking Milkey to establish that the harm Massachusetts would suffer is "imminent," a crucial element of standing. He demanded to know "when" the harm would happen, mockingly asking "When is the predicted cataclysm?"[15]

The question, and the tone of its delivery, left little doubt as to Scalia's disdain for scientific projections of catastrophic consequences from climate change. He then directly questioned the existence of a scientific consensus on climate change, distinguishing between whether climate change was happening, about which he seemed to acknowledge there might be a consensus, and whether it might be "attributable to human activity," about which he suggested there was "not a consensus."[16]

Before Milkey could fully tackle this issue, Scalia tried to pin him down on precisely how much greenhouse gas emissions from new motor vehicles in the United States were contributing to harmful effects caused by climate change "right now." What, Scalia demanded to know, was the precise percentage of greenhouse gas emissions coming from those new motor vehicles and what was their worldwide significance? How did those emissions establish "how remediable the harm is"—another crucial Article III standing requirement?[17]

When Milkey acknowledged that new motor vehicle emissions accounted for 6 percent of worldwide emissions, Scalia probed further in rapid-fire fashion to demonstrate that the petitioners had failed to meet the "remediable" requirement. Even assuming there was current harm that might somehow be attributable to current greenhouse gas emissions, how much, if any, of that harm would be redressed by decreasing that one source of emissions by a marginal amount? After all, Scalia stressed, new motor vehicle emissions were only a small percentage of total U.S. emissions, which were an increasingly small percentage of worldwide emissions. Further, any EPA restriction on new motor vehicle emissions would not reduce that already-small U.S. percentage to zero; it would just reduce those emissions at the margin. The EPA could therefore, Scalia concluded, at most only decrease greenhouse gas emissions by a small percentage over time, with very little impact on global emissions levels—"the 6 percent will only be reduced to maybe five and a half in the next few years."[18]

There was legal force to all of Scalia's questions on both the "imminent" and "remediable" requirements for Article III standing. Although some harm from climate change occurs "right now," that harm is not a direct result of current greenhouse gas emissions. It results from past greenhouse gas emissions that have been settling in the atmosphere over decades—even centuries—leading to higher and higher concentrations. Greenhouse gas emissions today will cause harm not in the next days or months, but in the next decades or even centuries. That hardly smacks of "imminent harm," something the Supreme Court says is required for a plaintiff to be entitled to sue for relief in federal court.

Compounding the imminence problem, Scalia's further point—that EPA regulation of emissions from new motor vehicles in the United States would have only an exceedingly small effect on total emissions from all sources everywhere—also seemed damning to Massachusetts's claim of standing. How could Massachusetts establish that any such reduction—small from a global perspective—would prevent particularized, identifiable harm to the Commonwealth or to any of the other petitioners from climate change in the future, let alone imminently?

There is no ready scientific formula for calculating what level of reduction of greenhouse gas emissions will avoid specific harms in a particular place at a particular time. Besides, emissions reductions in one part of the world could rapidly be overwhelmed by emissions increases in other parts. The enormous temporal and spatial dimensions of climate change—spanning decades, if not centuries, and circling the globe—resist ready application of the Supreme Court's requirement that plaintiffs demonstrate standing by establishing how their requested relief will "redress" "concrete, imminent harm."

Scalia's questions were smart. They presented Milkey with rhetorical traps to navigate, and the verbally dexterous Justice was ready to seize on the slightest hint of a misstep. No Justice was better than Scalia at demonstrating with gusto how a lawyer's answer proved too little, too much, or nothing at all.

Milkey had seen his case fall apart when he had faced questioners playing the role of Justice Scalia in the practice moot court arguments. They had been able to control the argument and prevent him from getting across his most important points. He had practiced parrying back and had improved his verbal footwork, but in none of those practice sessions had Milkey faced the person now looming large only a few feet before him.

Anticipating precisely this challenge, Milkey had jotted down a few cryptic notes in longhand earlier that morning, which he had brought into the courtroom and placed on top of all his other notes. They were simple, direct admonishments to himself:

—no wind-up or filler
—stay on track
—get back on track
—answer the ? asked
—eye contact
—time = your enemy[19]

Standing before the actual Justices, Milkey clicked into high gear. The failed moot courts were history. Milkey parried every one of Scalia's twenty-three questions with direct, succinct, and responsive answers. He maintained a clear thesis and stuck to it in a respectful, credible, and persuasive manner. The perpetual second-guessing, the torrent of skeptical questions during the moot courts, and seemingly endless debates within the Carbon Dioxide Warriors about what they should and should not argue had given him the knowledge he needed now to speak directly and with conviction. And the brutal moot courts had steeled him for the onslaught of pointed questions, more than in any other case he had argued in his decades as a lawyer.

Milkey addressed the question of imminence by pointing to the affidavits submitted by Massachusetts alleging that the rises in "sea levels are already occurring" and "it is only going to get worse" with increased emissions. He calmly challenged Scalia's mocking reference to the "predicted cataclysm" by soberly explaining that their claim of standing did not turn on such extreme, dire predictions only in the future. The harm the state alleged was "ongoing harm" that "plays out continuously over time. . . . Once those gases are emitted . . . they stay a long time, the laws of physics take over." There was nothing "conjectural" about it. Nor would it "suddenly spring up in the year 2100."[20]

Milkey bolstered his argument that their requested relief was necessary to prevent further harm from climate change by pointing out that although this case involved only emissions from new motor vehicles, the EPA had "disclaimed authority to regulate all sources [of carbon dioxide]" and not just those from automobiles. Justice Ginsburg quickly picked up this point and underscored it. She helpfully explained that if the EPA were correct about the meaning of the Clean Air Act, the EPA would lack authority to reduce greenhouse gas emissions of all sources, including "carbon dioxide in power plants."[21] Scalia's opening barrage on standing ceased.

Milkey's confidence was palpable as he realized he had expertly parried Scalia's questions. And he was equally effective when Chief Justice Roberts and Justice Alito joined in the questioning. When the Chief suggested that Massachusetts's claim of injury was, like in the Court's cases denying standing to taxpayers who were upset that their taxes funded governmental programs they opposed, too general in character, too "small," and too "widely dispersed" to support standing, Milkey quickly distinguished his case from the Court's earlier taxpayer standing decisions by asserting that "it is different because here there is *particularized injury* that we have shown"— meaning that the injury to Massachusetts from climate change was both severe and specific to the Commonwealth. Milkey then brought that point home by emphasizing the sovereign nature of the injury Massachusetts would suffer from climate change: "The injury doesn't get any more particular than states losing 200 miles of coastline, both sovereign territory and property we actually own, to rising seas."[22]

Justice Alito took his shots but was no more effective. He followed up on Scalia's skepticism that reductions in new motor vehicle emissions would reduce Massachusetts's climate injury sufficiently to allow for standing in federal court. From a global perspective, Alito stressed, the reductions achieved would be "small" "under the best of circumstances."[23]

Once again, Milkey navigated the question, shielding his argument from its weaknesses with strong, direct responses. He provided the precise percentages requested by Alito and forcefully made it clear that "given the nature of the harms, even small reductions can be significant. For example, he noted, "if we're able to save only a small fraction of the hundreds of millions of dollars that Massachusetts park agencies are projected to lose, that reduction is itself significant."[24]

It was a bull's-eye answer, well targeted to Justice Kennedy even if nominally offered in response to Alito's question. Milkey had pointed to the most traditional type of legal injury: physical damage to land. Property rights in land, Milkey knew, were of special concern to Kennedy. Milkey also scored a point with Kennedy by emphasizing that Massachusetts would suffer climate change injuries in its sovereign capacity. Climate change would literally reduce the borders of the state—a harm that would cost Massachusetts millions of dollars to mitigate.

Milkey's oral argument was significantly bolstered by his repeated references to affidavits filed by state officials in support of his claim that climate change would harm the Commonwealth of Massachusetts in significant ways. Although the Democratic attorney general, Tom Reilly, had litigated the *Massachusetts* case without any involvement from the state's Republican governor, Mitt Romney, Romney's general support early in his administration for addressing climate change allowed state agencies under his jurisdiction to cooperate with the attorney general's litigation by providing Milkey with affidavits that declared how climate change would literally erode the Commonwealth's borders and increase administrative costs for its parks.[25]

Justice Scalia tried one last time to land a clean blow by questioning Milkey on the validity of his apparent assumption that there was a "straight line ratio" between reductions in greenhouse gas emissions and reductions in climate change harms.[26] Milkey's response was the Supreme Court advocate's equivalent of the boxer's knockout punch. He turned the momentum of Scalia's attack right back at him, landing a hit in Massachusetts's favor.

Milkey clarified that he was not assuming a linear relationship but that, in fact, the lack of linearity meant that Massachusetts might well be suffering *more,* not *less,* harm. He backed this up with specific examples for both Massachusetts and New York, examples that were easily accessible to nonscientists like the Justices and in the case record. And he wrapped it all up by tying these specific examples to his broader legal point about the existence of current harm:

> Your Honor, I don't believe it's established it's necessarily a straight line. But I want to emphasize that small vertical rises cause a large loss of horizontal land. For example, where the slope is less than 2 percent, which is true of much of the Massachusetts coastline, every foot rise will create a loss of more than 50 feet of horizontal land. And for example, in the State of New York, the Oppenheimer affidavit projects that New York could well lose thousands of acres of its sovereign territory by the year 2020. So the harm is already occurring. It is ongoing and it will happen well into the future.[27]

If there were any remaining doubt that Milkey had effectively shot down the best punch that Justice Scalia could offer, it was erased a few seconds later when Justice Kennedy spoke. The significance of that moment was not lost on anyone in the courtroom. Anything Kennedy said would open a window into his thinking and, very likely, how the Court would rule.

Kennedy asked Milkey whether he believed that states possess some "special standing" and if he had a case that supported that argument. Milkey said they did and discussed a federal appellate court ruling that supported this idea. Kennedy then responded with every advocate's worse nightmare: a case they don't know. According to Kennedy, Milkey's "best case" was the Supreme Court's 1907 decision in *Georgia v. Tennessee Copper*.[28] None of the dozens of briefs had cited the case. None of the lawyers at the moot courts had mentioned it. Luckily, the Justice did not ask Milkey a direct question about the case. And so, Milkey quietly thought to himself, if this *Georgia* case "is good enough for Justice Kennedy, it's good enough for me."[29] Milkey had reason to be happy: It was clear both that Kennedy had done some independent research and that he believed the Court's precedent favored Massachusetts.

Not one to give up easily, Scalia tried to land one final punch on the standing issue. He pushed Milkey once again on whether the harm could really be considered "imminent," given how many decades would pass between the time greenhouse gases were emitted into the atmosphere and the time when those additional greenhouse gases would cause harm. This time, Milkey responded with a succinct zinger: "Your Honor, once these are emitted the laws of physics take over, so our harm is imminent in the sense that lighting a fuse on a bomb is imminent harm." With a simple, everyday example, Milkey had explained how a threat could simultaneously be both in the present and in the future.[30]

All the lawyers in the courtroom could see what had just happened. Massachusetts could count to five on the standing issue.

"Troposphere, Whatever."

Looking at his watch and thinking to himself "holy shit, we are nineteen minutes in and I haven't even gotten to the second issue"[31]—whether greenhouse gases are air pollutants—Milkey announced his attention to do just that: "Your Honor, if I may turn to the merits quickly."[32] But as soon as he tried to pivot to his argument that greenhouse gases are air pollutants under the Clean Air Act, the

Chief Justice steered Milkey away: "Moving from your authority argument to the exercise of authority . . ."[33]

The reason for the Chief's deft maneuver was plain to Milkey. Roberts had apparently concluded that Massachusetts was headed to a possible win on the question of the EPA's authority under the Clean Air Act to regulate greenhouse gases. For that reason, the Chief saw little value in spending scarce argument time on the issue. And, for strategic reasons, the Chief preferred to focus the Court's attention on the final issue of the case—whether the EPA had validly exercised its discretion when it had chosen to defer its decision on whether to regulate greenhouse gas emissions—an issue for which Massachusetts's arguments were far weaker.

The Chief pressed on the timing of the EPA's decision. He asked Milkey to tell him precisely what constituted the impermissible ground on which the EPA had relied in declining to decide whether to regulate greenhouse gases. Was there a statutory deadline that the EPA had missed? Or anything in the statute that otherwise dictated when the EPA must decide? When exactly had the EPA taken too long? After "day one"? Couldn't the agency validly assert that it wanted to postpone further decisions on the "principle that they want to deal with what they regard as the more serious threats sooner"?[34]

It was a classic Roberts series of questions. Not belittling, as Scalia's questions could be. But demandingly focused and insistent. Before becoming a judge, Roberts had himself been a celebrated Supreme Court advocate, widely considered one of the most gifted in the nation. He knew all the tricks of the trade and the vulnerabilities of different types of legal arguments, many of which he had once himself delivered in that same courtroom in front of the Justices who were now his colleagues. Roberts was especially adept at challenging counsel to identify a limiting principle to their argument, knowing full well that they are often very hard to pinpoint with any precision. Here, that meant asking Milkey to identify the precise moment in time when the EPA would be required to make an endangerment determination, and then explain the source of law that backed up that view.

On this issue, Milkey had a problem. There was nothing anywhere in the hundreds of pages of the Clean Air Act that imposed a deadline on the EPA to decide whether regulation was warranted. There was also no doubt the agency had vast discretion to decide to postpone a decision on greenhouse gases based on any of a zillion valid reasons. So he gave the only answer he fairly could. It was far from an obvious winner, but it was the only plausible argument available to him. He acknowledged that the EPA could have offered, as the Chief had suggested, legitimate reasons for delaying a decision. But, he added, it hadn't. Put simply, Milkey argued that the EPA had blown it because they "do not rely on any of those grounds, they do not rely on lack of information, they did not rely on background principles of administrative law."[35]

Instead, Milkey stressed, the EPA relied on an impermissible ground, and he pointed to the precise two sentences in the record of the case where the agency had done so. In "two back to back sentences on page A-82 of the third petition," the agency had made it clear that it rejected the petition to regulate because its preferred "policy approach" was "different" from the one Congress had required—to limit the emissions of any pollutant that endangered public health and welfare—even if the Bush administration thought it made more sense to try voluntary, nonregulatory greenhouse gas programs first. But, Milkey concluded, "rejecting mandatory motor vehicle regulation as a bad idea is simply not a policy choice Congress left to EPA."[36]

Ginsburg then jumped in to make sure that Milkey understood the import of this argument: If Massachusetts won on that narrow ground, the EPA could go back and reach the same decision on remand, based instead on one of the many reasons for delay that Milkey had acknowledged were legitimate. In response, Milkey acknowledged that under "background administrative law principles" the EPA could say something like "We just don't want to spend the resources on this problem and we want to look elsewhere" and Massachusetts would be very limited in its ability to challenge that determination. But, Milkey asserted, it had not yet actually said that. "It's important that EPA say that," Milkey insisted, and until they do, the agency had acted unlawfully.[37]

Of course, Milkey would have preferred to claim that the EPA would not be able to do any such thing on remand. The problem was that this would almost certainly have been a losing legal argument. He would do his client no favor by proffering an argument that lacked merit. His only viable option was to offer the only plausible argument he had—that the EPA had offered an impermissible reason for declining—and hope it might be enough to garner a majority.

Frustrated by the traction that Milkey's arguments seemed to be getting, Scalia returned to the "authority" issue the Chief had deftly sought to avoid. But here the Justice stumbled, leading to the argument's one lighthearted moment. Scalia expressed great skepticism about the notion that an "air pollutant," within the meaning of the Clean Air Act, could include a "stratospheric pollutant." After all, the Justice admonished, such a pollutant does not endanger what we "normally call 'air'" and instead goes up to the stratosphere, where it contributes to global warming.[38]

Scalia's contention was doubly flawed, as Milkey quickly pointed out. Most fundamentally, the Clean Air Act's definition of air pollutant required that it be emitted into the "ambient air," and greenhouse gases easily meet that criterion. Further, as Milkey explained, "there is nothing in the Act that actually requires the harm to occur in the ambient air." Milkey's point was that whether a chemical compound is an "air pollutant" turns on where it is emitted in the first instance—into the ambient air—and not precisely where in the atmosphere that compound causes harm.[39]

Milkey backed up his point by referring to the damage caused by sulfur dioxide in acid rain. Sulfur dioxide is emitted into the air, but causes harm only when it lands on trees and buildings as acid rain. The same could be said for many other classic air pollutants. For example, lead is emitted into the air but causes harm only when it subsequently lands on food, which is then ingested by people, resulting in dangerous concentrations of lead in the bloodstream.[40]

Milkey could not stop himself from also correcting Scalia on a scientific mistake: "Respectfully, Your Honor, it is not the stratosphere. It's the troposphere."[41]

Scalia's self-deprecating reply caused a ripple of laughter throughout the courtroom: "Troposphere, whatever. I told you before I'm not a scientist." But it was his next statement that, while tinged with humor, may have been the more revealing: "That's why I don't want to have to deal with global warming, to tell you the truth."[42]

The Justice did not believe that climate change lawsuits like the one brought by Massachusetts should be in federal court at all. It was, however, becoming clear that Scalia might be in the minority in this view.

With three minutes remaining, Joe Mendelson, who had been seated next to Milkey at counsel table, signaled to Milkey with a quick cough that he needed to reserve time for a rebuttal.[43] It was one of the few things Milkey had asked his co-counsel to do. The cough was critical. It is easy for an arguing counsel, in the midst of nonstop questioning from the Justices, to forget to save time for a rebuttal after the opposing counsel has completed his or her argument.

Adeptly cued by Mendelson, Milkey promptly reserved his remaining time for rebuttal and sat down. He felt good, and relieved. But he knew he was not yet done.

Signaling that it was now time for the federal government's attorney to argue, the Chief Justice simply stated, "Mr. Garre."[44]

16

Red Lights

Because *Massachusetts v. EPA* was so high-profile and politically controversial, it was the kind of Supreme Court case that normally would have been argued by the solicitor general, Paul Clement, himself. But in the same two-week argument session Clement had to argue a civil rights dispute that was even higher profile than *Massachusetts:* two consolidated cases about affirmative action that considered the legality of integrating public high schools through admissions programs that accounted for a student's race.

So, with Clement unavailable, *Massachusetts* became the responsibility of the principal deputy solicitor general, Greg Garre. Garre and Clement were the only two attorneys in the Solicitor General's Office who were political appointees. The career attorneys in the office would have had no problem arguing the case, but its political dimensions made it more suitable for a political appointee.

Garre was an experienced Supreme Court advocate. Just shy of six feet tall, broad-shouldered and bespectacled, with close-cropped thinning sandy hair, Garre combined a friendly and understated demeanor with a strong authoritative voice. He was a graduate of Dartmouth College and George Washington University School of

Law, where he was a classmate of Joe Mendelson, although they did not know each other there. Only forty-two, seven years Milkey's junior, Garre had already argued seventeen times before the Court, including three times that calendar year. He knew the Justices and they knew him. And before 10 a.m. that morning he had good reason to be confident that the EPA would prevail.

Like Chief Justice Roberts, Garre had clerked for Chief Justice William Rehnquist. And immediately following that clerkship, Roberts had hired Garre to work for him at a prominent D.C. law firm, Hogan & Hartson, where Roberts headed that firm's Supreme Court practice. They worked closely together there for seven years, and Roberts had been Garre's primary professional mentor and was still a good friend. In the month when Roberts was sworn in as Chief Justice, Garre was appointed principal deputy solicitor general. A few weeks earlier Roberts and Garre had served as pallbearers at the funeral of their former boss, Chief Justice Rehnquist.[1]

Rehnquist was beloved by his law clerks. His brilliance as a lawyer was clear, but it was his striking modesty, commonsense reasoning, and self-effacing personality that charmed them all. Even though he had himself graduated first in his class at Stanford Law School, he was far more willing than any of his colleagues to hire from law schools beyond the nation's so-called elite schools, believing that the top students at any of fifty or more law schools could do the law clerk job. His hiring of Garre, who had excelled at George Washington Law School, reflected that view. He preferred to discuss cases with his clerks by taking strolls outside the Supreme Court building, rather than asking them to prepare lengthy bench memoranda. During those strolls, when stopped by tourists asking if he would take their picture in front of the Court, he would cheerfully agree, leaving them wholly unaware their picture had been taken by a Justice. He loved to discuss history and geography with his clerks, and delighted in playing competitive games of all sorts, willing to wager small amounts on most anything. Rehnquist strongly encouraged his clerks not to measure their life by their professional achievements, but instead, as Garre fondly recalled, by "the 'fruits of life'—whether that's spending time with family, spending weekends riding your bike, or pursuing some other interest." Rehnquist once proudly declined to attend a State of the Union Address because it fell on the same night as his painting class at a local YMCA.[2]

In arguing before the Court, Garre found Chief Justice Roberts's familiar face comforting, but he always understood that he was arguing before the Chief Justice, not his friend, former colleague, and fellow Rehnquist clerk. Garre knew that he received no favor because of their long-standing relationship. While, to outsiders, the Chief might seem harsh and especially demanding in his questioning of Garre during oral argument, Garre understood that it was a sign of the Chief's respect, professionalism, and high expectations.[3]

As principal deputy solicitor general, Garre enjoyed the prestige and perks of his office, which is responsible for all Supreme Court litigation on behalf of the United States, some 80 percent of the cases decided by the Court. The Justices accord heightened weight to the oral arguments of attorneys from the Solicitor General's Office because they speak for the "United States"—so much so that the solicitor general is sometimes described as "The Tenth Justice."[4]

The Office's exalted role is underscored by several long-standing traditions. Within the Supreme Court building itself, just a few feet from the courtroom, there is a formal, private office with "Office of the Solicitor General" emblazoned on the door.[5] It's the equivalent of a locker room for the home team, with the added advantage that the visitors don't get their own locker room. Attorneys from the Solicitor General's Office even enter the courtroom differently, via the side entrance. And Court personnel reserve seats for them in the Supreme Court Bar section when they come to hear oral arguments.

Solicitor General's Office attorneys even have a home team uniform. By custom, they wear formal morning coats, including striped pants, to argue before the Court; all other attorneys wear normal business attire. The morning coat tradition has created challenges for the office as women have entered the ranks—some of whom have fashioned their own morning coat equivalents, while others, including former solicitor general and now Justice Elena Kagan, have drawn the attention of legal and fashion blogs alike by choosing to wear a suit instead when arguing.[6]

Fashion is not the only peculiar tradition of the office—again by custom, its attorneys argue *only* from the right side of the courtroom, closest to the door through which they enter. There are two tables at either side of the lectern, facing the bench, with assigned seating. The attorneys for the petitioner sit on the left side of the lectern and those for the respondent sit on the right—unless an attorney from the Solicitor General's Office is presenting an argument in support of the petitioner, in which case the counsel for the petitioner must switch to the right and the counsel for the respondent moves to the left.

All of these privileges and curiosities could well be seen as wholly beside the point. They don't change the votes that either side must obtain to win a case. But they do send a clear message. Some attorneys appearing before the Justices are insiders, others are outsiders. Garre was an insider.

"All of These Considera-SHUNS"

As he stepped up to the lectern, Garre wore an antique watch that his grandfather had given his father, and his father had in turn given to him. The watch had stopped keeping time years before, but that wasn't its purpose that morning: the family heirloom served as a good luck charm for oral arguments. On his drive to the Court from his home in northern Virginia, Garre had seen people in shorts and T-shirts out on the Mall stretching between the Capitol and the Lincoln Memorial, an atypical sight for a late-November day, giving him a "sinking feeling" that the warm weather was not a particularly good omen for the "global warming" case.[7]

When Garre stood before the lectern, he placed his argument folder before him—although, like Milkey, he knew he was unlikely ever to consult its contents. On the front of the folder, Garre had written the words "Enjoy" and "Be Thankful" just as he had done in all of his prior Supreme Court arguments. The words were there to remind him—notwithstanding the inevitable fatigue and stress of preparing and then presenting a Supreme Court argument—never to forget the enormous privilege and opportunity such an argument represented, especially when representing the United States.[8]

Garre knew that his argument was not a "slam dunk," but believed he had highly persuasive arguments on all three issues before the Court. He believed he was absolutely right on the standing issue, had strong Supreme Court precedent in favor of his argument that greenhouse gases were not Clean Air Act air pollutants, and was clearly right on the question of whether the EPA had the discretion to postpone any decision about whether greenhouse gas emissions endanger public health or welfare.[9]

Garre knew that an oral argument can make a big difference to the outcome of a case, and he prepared accordingly. He had spent approximately one hundred hours getting ready for the oral argument in addition to the time he had devoted to the written brief. He had identified likely questions and crafted crisp answers. He had also developed a series of short "blurbs" he would be ready to say during the argument to emphasize the themes he thought were most important for the Court to understand. He had had even met with scientists from the National Oceanic and Atmospheric Administration (NOAA) so he could better understand the science of climate change.[10]

Garre had a pretty good idea which Justices would prove the most challenging that morning. As he anticipated, those who had dominated the first part of the argument, seeking to derail Milkey, suddenly went largely silent, while those who had said the least during Milkey's argument rediscovered their voices to aggressively challenge

Garre. The dramatic shift invariably undercuts the appearance of strict neutrality the Justices otherwise strive to project to the public.

Garre's first three questioners were Justices Ginsburg, Breyer, and Stevens. During Milkey's argument, Breyer and Stevens had been completely silent and Ginsburg had spoken up not so much to question Milkey as to bolster his answers to Justice Scalia's questions. Now the Chief Justice would play that role, helping Garre out several times. Justice Thomas, who had asked no questions of Milkey, would do the same with Garre, consistent with his long-standing practice of almost never asking a question at oral argument—at the time of the *Massachusetts* argument, Thomas had asked questions in only two cases in the past twelve months and would go on not to ask a question in a single case until February 2016. The Justice has explained his silence during argument, which contrasts with his engaging personality and booming voice off the bench, by stating that his colleagues already ask too many questions and the advocates need more time to make their points.[11]

Garre began his argument by focusing principally on the third issue: whether the EPA had the discretion to postpone its decision on whether greenhouse gases from motor vehicles endanger public health or welfare.[12] His tactic was a product of his long experience arguing before the Court as counsel for the respondent, arguing second after counsel for petitioners. Good counsel for respondents don't rely on a text prepared beforehand. They are ready to craft arguments on the spot based on what was just said: which Justices asked what questions and how they reacted to the answers given.

Garre's beginning made it clear that he had concluded from Milkey's argument that the EPA's best shot at winning the case was on the third issue. He accordingly stressed that even if the Court concluded that the EPA possessed the authority to regulate greenhouse gases, the agency's decision that "now is not the time to exercise such authority" is a "quintessential administrative judgment" that the courts had no grounds to second-guess. Garre did not mention standing at all in his opening statement, and he gave only a passing nod to the argument that the EPA lacked authority over greenhouse gases.[13]

The second Justice to challenge Garre, Justice Breyer, immediately identified the weakest spot in Garre's argument. Breyer is known for his analytical rigor and careful attention to every detail in the record. He taught administrative law for decades at Harvard Law School, so he knew as much as anyone about the area of law upon which Garre was now relying. But once a law professor, always a law professor, and Breyer was known on the Court for posing questions that seemed better suited for a classroom than for the tight time constraints of a Supreme Court argument. In contrast to the kinds of short, succinct questions posed by many of his colleagues, Breyer's questions were typically lengthy statements replete with multiple clauses followed by an open-ended question such as "What do you think?" Advocates frequently struggle to identify the focus of Breyer's inquiry.

In *Massachusetts,* Breyer's question concerned the precise language the EPA had used in its decision to deny Mendelson's petition. The agency had denied the petition on the ground that, even if the EPA possessed regulatory authority over greenhouse gases, it was the wrong time to decide whether to exercise that authority. Breyer made one of his classic (very) long statements, followed by a short probing question at the very end:

> On this particular issue, the opinion as I read it, of the EPA, consists of 32 pages. Twenty of these pages, 22 in fact, deal with whether they have statutory authority. And of the 10 that deal with the issue we're talking about now, five of them give as their reason that they think that the President has a different policy. Of the remaining five, two more consider international aspects of the problem and how you have to get other countries to cooperate; and then the conclusion of that

part says in light of these considerations, we decide not to exercise our power.

Now their [Massachusetts's] claim with respect to that, is that at least three of the four considerations are not proper things for the agency to take into account: namely whether the President wants to do something different, whether we're running foreign policy properly, whether cooperation with other countries are relevant to this particular issue.

So what they've asked us to do is send it back so they can get the right reasons [n]ow—if they want not to do it. What's your response to that?[14]

Breyer's question plainly (and understandably) struck Garre as odd in its framing. Garre responded that the number of pages the EPA had given to different considerations in deciding to defer resolving whether greenhouse gases endangered public health and welfare was entirely beside the point. All that mattered, he stressed, are the "reasons the agency gave" and so long as at least one of those "considerations" is valid, the Court can and should sustain the agency's decision. He emphasized that it should make no difference to the outcome whether the Court believed that some of "the other considerations were inappropriate."[15]

Breyer's question, however, had laid a trap in its wording, burying his real point. And Garre had walked right into that trap by focusing on the number of pages, which was a red herring.

Breyer's true focus was not on page numbers but on the EPA's choice of words. The EPA had stated "in light of these considerations, we decide not to exercise our power." Breyer's point, best revealed by listening to the audio recording, is that the EPA had not concluded that its decision was supported by any one of those "considerations," on its own, but that it had found its conclusion necessary in light of all the "considerations." Breyer underscored this distinction through his repeated emphasis of the final "s" in "considerations." As one can hear clearly in the publicly available audio recording of the argument, Breyer's pronunciations of "considerations" became "considera-**SHUNS**."[16]

Breyer then went in for the kill:

> When I write an opinion, sometimes I write the words: "We decide this matter in light of the following three factors taken together." And I guess a lawyer who said, "one of those factors alone the Court has held justified the result all by itself"—in saying the Court has held that, I guess that wouldn't be so. That would be a bad lawyer, wouldn't it?
>
> . . . .
>
> If [EPA officials] write that all of these considera[-**SHUNS**] justify our result, again, one of them by themselves, it sounds, they think would not have been sufficient.[17]

Breyer's argument was that Garre's defense of the EPA's decision suffered from a fatal flaw. Garre was arguing that the decision could be upheld so long as any one of the many reasons the EPA had given in support of its decision was valid. But the EPA, by Breyer's reading of the agency decision, had never claimed that any one of those reasons, standing alone, would suffice. Just the opposite. It had relied on them all, meaning that if any one of them was not sound, the agency's decision must fail. Garre's argument therefore failed because it rested on a mischaracterization of the EPA's decision. For Breyer, this was the argument of "a bad lawyer."

Of course, no one, including Breyer, thought Garre was a bad lawyer. He was clearly an outstanding lawyer. The problem was that his client—prompted by political appointees in the agency—had committed a misstep in the precise words it had used in denying Mendelson's petition. Garre was now stuck trying to defend that unfortunate, or downright sloppy, language.

Milkey, Mendelson, and Heinzerling at counsel table and all the Carbon Dioxide Warriors scattered around the courtroom or listening to the audio from the lawyer's lounge had good reason to be elated by Justice Breyer's line of questioning. They had only one possible, very narrow opportunity to win against the EPA's final backup argument that the agency had discretion not to decide the endangerment issue at all. Breyer's embrace of that potentially winning argument did not, of course, mean that they would win on this issue, but it was about as good a sign as they could have hoped for that they might.

A few minutes later, Justices Stevens, Kennedy, and Souter posed questions that seemed to evince that they were concerned with Garre's suggestion that there was little to no limit on how long the EPA could decide to delay its decision. Their questions were the other side of the coin of the questions that the Chief Justice had earlier posed to Milkey. The Chief had asked Milkey to tell him exactly how little time the EPA had to respond to a petition—and now several other Justices were asking Garre to tell them whether there were any limits on how long EPA could delay an answer to a petition. Stevens asked whether the EPA had the "discretion never to make a judgment." Kennedy referred to precedent that limited an agency's ability to delay for an unreasonable amount of time. And Souter closed by suggesting, precisely as Massachusetts had argued, that it was not a "legitimate concern" for the EPA to decide not to make an endangerment finding because the EPA didn't want to exercise the authority Congress had given the agency "under the statute" if such an endangerment exists.[18]

Souter's was the last question asked. The red light on top of the lectern lit up, indicating that Garre's time was up. He could only respond by reiterating the EPA's position that "Congress had not authorized it to undertake the regulation of greenhouse gas emissions to address global climate change and that, even if it had, that authority should not be exercised."[19] And then he sat down.

An hour earlier, no one really knew how the Court might rule. But now, Massachusetts seemed to be on the edge of not just of winning, but of winning big. On all three issues.

Justice Scalia's battering of Milkey with questions had clearly advertised his hostility to the legal arguments of the *Massachusetts* petitioners. But he was not the only Justice who knew how to use oral argument to promote his preferred outcome in a case. Following Scalia's lead, many of the Justices did the same, and consequently the questions they asked provided tea leaves for predicting their votes and likely outcome.

As Garre took his seat, Milkey quickly ran through in his head the questions asked by the Justices over the past fifty-seven minutes. One (Ginsburg) . . . two (Souter) . . . three (Breyer) . . . four (Stevens) . . . five (Kennedy). He could count five possible Justices on his side. All he had to do was get through the last three minutes of argument—the amount of time he had reserved for rebuttal—without losing ground. Milkey could, at that point, have waived his rebuttal time, but he was not that confident of a win as he stood up and returned to the lectern to face the Justices.

"The Case Is Submitted"

Justice Scalia could see where the case was going, and he wasted no time in pouncing. Scalia's challenge was to quickly identify a way to turn the case back in his favor. Even before Milkey was able to settle at the lectern and say a single word, Scalia went to work:

> Mr. Milkey, do you want us to send this case back to the EPA to ask them whether if only the last two pages of their opinion were given as a reason that would suffice? Would that make you happy?[20]

Scalia's strategy was clear. Like Justice Breyer, in questioning Greg Garre, he was laying a trap. In the guise of an informal, almost flippant inquiry—"Would that make you happy?"—Scalia sought to entice Milkey into abandoning what had proven to be the central strength of his argument on the vital third issue before the Court.

Scalia had seen how Milkey had bolstered his argument by clarifying that Massachusetts was not asking the Court to do anything extraordinary. Massachusetts was not asking the Court to order the EPA to regulate greenhouse gases. Instead, it was merely asking the Court to remand the case back to the EPA in order to allow the agency to develop a better reason, if it could, for postponing its decision on whether to regulate greenhouse gases.

Scalia's question was designed to get Milkey to concede that Massachusetts in fact wanted more than it was admitting. That the state's true position was far more demanding than a mere request for a benign order sending the case back to the EPA for a further explanation of the reasons for its lack of action.

For a few seconds, the strategy worked. Perhaps because he was simply tired. Or perhaps because the significance of the question was well masked, Milkey took the bait. He quickly acknowledged, "It would not make us happy, your Honor." Justice Scalia embraced that answer by immediately replying, "I didn't think so."[21]

The courtroom filled once again with laughter—Scalia was famous for his wit and repartee and every year generated far more courtroom laughter than any other Justice—but only because most members of the audience were not aware of what Justice Scalia had just accomplished. They mistakenly assumed that the Justice was just being funny. But those who understood what was going on bristled in their seats with alarm. Milkey's answer had unwittingly provided Scalia with possible ammunition when the Justices later met to deliberate about the case. Scalia could now argue, with Kennedy in mind, that the import of Massachusetts's legal argument was far more extreme than the state had let on. The *Massachusetts* petitioners didn't just want to send the case back to the EPA—they were demanding greenhouse gas regulations.

What might have happened during the Court's deliberations if Milkey's response had ended then cannot be known, however, because the significance of Scalia's maneuver did not escape the attention of at least one person in that room who was paying close attention: Justice Breyer. Although Milkey tried to move on quickly to a point about standing, Breyer stopped him short. He was always on the lookout for efforts by his good friend Nino to throw a last monkey wrench into a case in an effort to prevent the Court from forming a majority consensus on a result that Scalia opposed (and Breyer favored). He brought Milkey back to the exchange with Justice Scalia. This time, there was nothing rambling or longwinded about his delivery. The remaining few seconds in Milkey's rebuttal time were quickly evaporating. Breyer needed to do damage control, and to help get Milkey back on track.

"What is your answer to Justice Scalia?" Breyer prodded. Then, without allowing Milkey to answer, he supplied Milkey with what his answer should be: "Because I thought you said before that you thought it was appropriate for us to send this case back so that [the EPA] could redetermine in light of proper considerations whether they wanted to exercise their authority. . . . Am I wrong about that?"[22]

Milkey immediately understood what Justice Breyer was doing, and what Scalia had been up to.[23] He grabbed the lifeline Breyer had thrown him: "Your Honor, that is exactly what we want."[24] Breyer had, in effect, out-Scalia'd Scalia—by rescuing a struggling advocate just as Scalia had done countless times before.

Scalia, followed by the Chief and Alito, then jumped back in to express disbelief that Milkey really thought the EPA wouldn't just go back and reach the same conclusion, based only on one consideration such as scientific uncertainty. This time, Milkey kept to his moorings. The EPA had not yet made that judgment, he insisted, and until the agency did, its decision to deny the petition remained unlawful.

It was 11:02. The red light on the lectern lit up, indicating that Milkey's time was up. Milkey formally closed by thanking the Court and sat down.

The first nine words of any Supreme Court oral argument—"Mr. Chief Justice, and may it please the Court"—are bookended by the last four stated by the Chief Justice after the allotted time for oral argument is complete: "The case is submitted."

At 11:02 the Chief Justice formally submitted the case, the Marshal banged the gavel, and everyone there for the *Massachusetts* case stood and exited as the Chief Justice and Associate Justices stood and stretched before the Chief called to the lectern counsel for petitioner in the second and last case of the day, a banking case.[25]

During the argument, Milkey's wife, Cathie, who had served as his "Justice" for the last moot court on the park bench the day before, was seated in the courtroom next to lobbyists from the auto industry. They could not help but give her a big smile and a "thumbs up" sign in recognition of the excellence of Milkey's argument.[26] They also privately confessed that they did not actually care so much about the outcome. "Either way we make money," the industry lobbyists joked. A sobering reminder of the ways of politics and business in the nation's capital.[27]

While leaving the courtroom, David Doniger, who had been Milkey's greatest skeptic and foe throughout the litigation process, told a colleague and Milkey himself that Doniger and the other Carbon Dioxide Warriors who had so strongly opposed Milkey presenting the oral argument were eating "crow-pie," given the terrific job Milkey had just done.[28] David Bookbinder, who had joined Doniger in trying to block Milkey from arguing and who that morning had challenged Milkey to prove that he (Bookbinder) was "the biggest asshole in the world," didn't acknowledge his new apparent status.

Milkey was elated as he left the building and walked down the courthouse steps with Cathie and his colleagues from the Massachusetts Attorney General's Office. The lead-up to the oral argument had been one of the most painful periods of his life. It had strained friendships, and ended a few. He has not spoken once since with

Heinzerling, Doniger, or Bookbinder and he did not even join them for what should have been a routine celebratory lunch that day. The case had forced him to draw on deep reserves of stamina. And now, with the bang of a gavel, it was done. As he reached the bottom of the steps, he felt like he "just had been released from prison."[29]

17

The Conference

Marshal Pamela Talkin banged her gavel. Everyone in the courtroom stood, and all eyes watched silently as the nine Justices slipped back behind the curtains. Unnoticed was the simultaneous exit of nine far younger women and men with unique responsibilities. They had watched the proceedings from the worst seats in the house, old rickety wooden chairs with partially obstructed views on the south side of the room. But unlike everyone else heading out through the massive oak doorways on the west side, which funneled them into the Alabama marble columned Great Hall, these nine left through a modest back door on the east side that led directly to the secure areas where they each worked: the Justices' chambers.

Each July, thirty-six recent law graduates arrive at the Supreme Court to begin their one-year clerkship with one of the nine Justices. They represent the top graduates of the nation's law schools, though many have not yet taken the bar exam and are therefore not even licensed lawyers. The four clerks hired by each Justice provide legal research and writing assistance: reviewing the petitions seeking the Court's review, reading the briefs filed on cases granted review, preparing memoranda on cases that are about to be argued, and assisting in the preparation of the Justice's written opinions.

It is an opportunity that can change the trajectory of any young lawyer's career. After completing their clerkships in 2007, the year when *Massachusetts* was decided, law firms offered these clerks a signing bonus of $200,000 on top of a base annual salary of $160,000.[1] By 2019 the bonuses had risen to $400,000, with a starting base salary of $190,000.[2] The Chief Justice of the United States, by contrast, earns $267,300 per year (just shy of $12,000 more than the Associate Justices) and the American taxpayers offer no signing bonuses.[3]

While working at the Court, the law clerks are deliberately anonymous. It is public knowledge that a Justice has hired them, but everything else about their clerkship is off limits. There is nothing subtle about the expectation that they will hold in strict confidence the workings of the Court and their chambers. There is a formal, written "Code of Conduct for Supreme Court Law Clerks" that unequivocally states that "a law clerk should never disclose to any person any confidential information received in the course of the law clerk's duties."[4] Clerks are told on their first day on the job that any breach of confidentiality will result in immediate termination.

The subset of nine law clerks sitting in the courtroom that day were not there accidentally. They had been primarily responsible for providing their Justices with assistance in the *Massachusetts* case. By custom, the Justices never decide which of their four clerks will work on any one case. They leave it to the clerks to work it out among themselves, with different clerks finding different cases more or less interesting. Justices not infrequently privately wish they could pick certain clerks for specific cases—based on their differing abilities and backgrounds—but remain neutral to avoid undermining morale and stirring up destructive rivalries within the chambers.

Clerks in some chambers use a selection system that is akin to the one most famously used by the National Football League for its annual draft of players, with multiple rounds of selections. Other chambers are less formal, believing in a "no-drama, laid back approach to things," a process that nonetheless allows clerks to choose their favorites.[5] The *Massachusetts* case was one of the more high-profile cases before the Court; each of the nine clerks witnessing argument would have ranked the case very highly and selected it in an early round or otherwise designated the case as a favorite.

Because the Court's confidentiality extends to which clerk worked on cases, the identity of the nine *Massachusetts* clerks is not known. But five of the thirty-six law clerks clerking for the Justices at the time had special reason to be interested in the case, having previously clerked for one of the three judges who had decided the case when it was before the D.C. Circuit. Two former clerks for Judge David Tatel were now clerking at the Court: one for Justice Stevens and another for Justice Souter. And three former clerks for Judge David Sentelle were now clerking for Chief Justice Roberts and Justice Thomas. By tradition, though, if the clerk had actually previously worked on the case when it was decided by the lower court, that same clerk would not work on the case at the Supreme Court precisely because of their past close relationship with the case.[6]

For most clerks, the single year with a Justice proves to be the preeminent professional experience of their lives. The clerks are trusted confidants of the Justice on the most sensitive legal issues facing the Court and the nation. Justices do not talk about the cases to anyone outside the Court, and they tend to be guarded in discussing their thinking with their fellow Justices, whose views may differ widely from their own. Their law clerks are the people with whom they work most closely and with whom they can have the most frank exchange of views.

It is a relationship that generally persists informally for years, even decades after the clerkship ends. Indeed, a clerk's association with their Justice is lifelong. Other lawyers will invariably refer to former Supreme Court clerks by the name of the Justice for whom they clerked—for instance, as a "Roberts clerk," "Stevens clerk," or "Scalia clerk." In 2016, when Justice Scalia's casket was somberly carried down the Supreme Court steps, his path was lined on either side by almost every one of the approximately 120 lawyers who had served as his law clerks. Some were then in their fifties, and had gone on to distinguished careers as a U.S. solicitor general, trial and appellate judges, legal scholars, and leading Supreme Court advocates. All stood respectfully at attention as their former boss left the Court for the last time, as did Justice Stevens's clerks for their Justice on those same steps in 2019.[7]

Fresh out of law school, clerks spend their days and nights (and weekends) working on the most important legal issues of the day in a building that serves as a constant reminder of the historic nature of their work. Unlike members of the public, they have full access to the Court's most secure areas, including what they refer to as the "Highest Court in the Land": the seventy-eight-foot-long basketball court on the top floor.[8]

There is, however, one room that no clerk can enter when the Justices are present—and that is the conference room, the building's most private room, located directly behind the courtroom within the Chief Justice's suite of offices. When the Justices gather there to meet, its three doors remain firmly shut as the Justices deliberate and vote on the outcome of each case—only the nine Justices are in the room.[9] It was in this conference room that the Justices would meet next on the morning, Friday, December 1, following the oral arguments in *Massachusetts v. EPA,* presented just two days earlier.

Breaking Bread

There is, of course, another gathering space for the Justices, and that is the Court's dining room. On most days the Justices dine alone in their chambers or go out to lunch with their law clerks. Sometimes, just for fun, they'll dine out with the clerks from another chamber.

But on argument days, they meet immediately after for lunch in the Court's dining room, a strikingly formal room where they are surrounded by history. The chairs, which were specially crafted in 1795, are carefully arranged. The room's centerpiece, a rectangular mahogany table, was created by a New York cabinetmaker in the early nineteenth century. Before the new building opened in 1935, the only place the Justices could eat together was the cramped quarters offered by their robing room, and the Justices often worked at home because they lacked individual judicial chambers.[10]

At these meals in the current building—which are entirely for socializing, not working—traditions abound. The Justices have assigned seats, determined this time not by seniority but by lineage. Each Justice sits in the spot where his or her immediate predecessor sat, and so forth back to the first Justice to be named on the Court to their seat.

So Chief Justice Roberts was in Rehnquist's seat, which had been Chief Justice Warren Burger's seat before him and Chief Justice Earl Warren's before that, tracing back to the Court's first Chief Justice, John Jay. Justice John Paul Stevens's immediate predecessor was Justice William Douglas, and the first holder of his seat was Justice John Blair Jr., who joined the Court in 1790. Justice Kennedy's seat could be traced back to Justice John Rutledge, who also joined the Court in 1790, and Justice Ruth Bader Ginsburg held the seat first held by Justice Thomas Todd in 1807.[11]

There were not always nine Justices. The number of Justices is determined by statute and not established by the Constitution. There were only six Justices when the Court first met in the late eighteenth century. During the nineteenth century, Congress changed the number of Justices six times, with the number dipping as low as five in 1801 and as high as ten in 1863, before settling at the current nine seats in 1869. President Franklin Delano Roosevelt famously sought in 1937 to increase the number of Justices on the Court— proposing in his "court-packing plan" to add up to six new Justices for every Justice over the age of seventy years and six months—but even the extremely popular Roosevelt was rebuffed by Congress, including members of his own party.

The Court's traditions and rituals, including the lineage-based seating arrangement, have nothing incidental about them. They underscore the strong ties of the Justices to their predecessors and, more critically, to the binding precedent reflected in the hundreds of cases decided by the Court long before they joined the bench. A climate change case hinging on future projections of damage from car emissions might seem wholly unrelated to the kind of legal issues the first Justices faced at the nation's founding. But those early cases guide the current Court's decision making with surprising frequency, and the high standards of analytic rigor and legal reasoning established by the Court two hundred years ago apply no less to its work today.

These well-worn traditions bind the Justices to the past, but the cordiality of sharing a meal binds them to each other. Like their predecessors, the Justices build bonds with good conversation. In contrast to his silence on the bench, Justice Thomas is garrulous, spirited, and engaged in the lunchroom, fully enjoying his colleagues. His frequent, booming laughter fills the room. He and Breyer have long enjoyed each other's company, and Justices Ginsburg and Scalia and their spouses were close friends for decades, sharing a love of opera and holiday dinners at their respective homes. Both proud midwesterners, Stevens and Roberts were always especially interested in news about Chicago. Not surprisingly, all the Justices are on a first-name basis, except that by tradition they each refer to the Chief Justice only as "Chief," even during lunchtime, a practice Scalia and Stevens together announced to Roberts on his first day at the Court. Decades his senior in both age and service on the Court, they cheerfully introduced themselves as Nino and John while telling him they would insist on calling him "Chief."

Favorite lunchtime topics of conversation included books, art exhibitions, concerts, and news about their respective families. Justice Kennedy is an avid reader and a serious student of history. Later in life he even created for his grandchildren his own curated catalogue of books, speeches, poetry, movies, plays, and songs that he captioned "Understanding Freedom's Heritage: How to Keep and Defend Liberty."[12] Justice Ginsburg's presence tended to discourage extended conversation about sports, as her lack of knowledge of sports was exceeded only by her lack of interest. Although discussions of argued cases were off limits, gossip about the attorneys who argued before them was strictly in bounds.[13] On November 29, the day that the *Massachusetts* case was argued, neither of the *Washington Post*'s front-page headlines—about the Iraq War or a newly elected Democrat senator's reluctance to be photographed with President Bush—likely provided fodder for lunchtime conservation. Nor was Justice Thomas, an avid reader of the *New York Post* sports pages, likely to have brought to his colleagues' attention the lead article that day on the Chicago Bulls beating the New York Knicks in basketball the night before.

To be sure, the Justices' dining rituals are a far cry from the Court's early days under Chief Justice John Marshall, when the Justices lived together in a boarding house. Their shared meals were famously "lubricated with well-chosen Madeira" (for which Marshall was well known), which "allowed the justices to exchange views frankly and to arrive at a common understanding on the points at issue."[14] Justice Joseph Story reportedly was a nondrinker when he first came to the Court but was soon persuaded by Marshall to join him for a glass of wine whenever it rained, which, he further pointed out, must be happening somewhere in the Court's large national jurisdiction on any given day.[15] Marshall brought to the meals a bottle with a "Supreme Court" label on it.[16] (Unconfirmed reports have suggested that Marshall was the largest importer of Madeira wine to the United States at the time.)

Justices nowadays do not tend to share the same food and drink. Diets differ in ways that reflect the Justices' distinct backgrounds, and even hint at their jurisprudence. On the day of the *Massachusetts* argument, New Englander Justice Souter would have enjoyed, as he did every day, a spartan lunch consisting solely of a plain yogurt. He saved a piece of fruit for an afternoon treat. Midwesterner Justice Stevens favored cheese or peanut butter sandwiches (sometimes with banana slices added to the peanut butter) with the crust carefully cut off: an everyday man's fare with a hint of the indulgent lifestyle that Stevens enjoyed as a young child growing up in a wealthy family in Chicago. Raised in Queens, Justice Scalia typically chose a heartier fare, such as pasta or a hamburger, perhaps proudly ordering the dessert avoided by the more weight-conscious Justices.[17]

Although the precise menu of the Justices has changed over time, the institutional purpose of the shared meals remains unchanged: to promote collegiality and remind the Justices, despite their differences, of their shared mission to maintain the integrity and independence of the Court. Given the claims of an often fractured world on the Court's attention, the power of a shared meal to forge ties has always been understood. Justice Thomas has publicly acknowledged the enormous institutional value of their shared meals, commenting that "it is hard to be angry or bitter with someone and break bread and look them in the eye."[18]

"I Will Do the Right Thing"

Two days after the oral argument, the Justices met in the conference room to discuss the case. Unlike their lunch on the afternoon of the argument itself, this meeting was not social. The Chief sometimes orders in for sweet rolls or cookies when he feels a break is needed, but other than that the conference meeting is all business.

The conference room is a beautiful room designed with a clear purpose, dominated by a three-foot-by-twelve-foot rectangular mahogany table with black leather inlay. Nine black leather-backed chairs surround it, with the name of a Justice engraved on a bronze metal plate on the back of each chair. A brilliant crystal chandelier hangs above the table, and portraits of a select few former Justices, including John Marshall and John Jay, line the northern Indiana white oak walls. The chair assignments are determined, as they were on the bench in the courtroom, by strict seniority and not lineage.[19]

On December 1, 2006, when the Justices met to discuss the *Massachusetts* case, Roberts sat in the seat assigned to the Chief, alone at the table's east end, and the next most senior Justice, John Paul Stevens, sat across from him. To the Chief's immediate right in increasing order of seniority sat Justice Souter, followed by Justices Kennedy and Scalia, closest to Stevens. To the Chief's immediate left sat the next four Justices in decreasing order of seniority: Justices Thomas, Ginsburg, Breyer, and Alito, closest to Justice Stevens.[20] As the junior Justice, Alito was responsible for answering any knock at the door. He was also the official note-taker, responsible for recording the votes of the Justices on all matters discussed that morning.[21]

Every Chief can choose which portraits to hang in the conference room and sometimes will pick a portrait deliberately to convey a message to his colleagues. William Rehnquist had chosen at one point to hang the portrait of Justice John McLean to the immediate left of Chief Justice Marshall's portrait. McLean had dissented from the Court's infamous *Dred Scott* decision, which had ruled on the eve of the Civil War that African Americans were not American citizens.[22] His portrait in the room stood as a reminder to the Justices of the stakes of their rulings, the Court's fallibility, the virtues of independence, and the ultimate judgment of history.[23]

Lining the conference room's west wall are built-in bookshelves with hundreds of volumes of the Court's prior decisions that, unless overruled, the Justices are bound to follow. On page 230 of volume 200 was the Court's century-old decision in *Georgia v. Tennessee Copper,* which Justice Kennedy had brought up during the oral argument, suggesting that it supported the Court's now finding that Massachusetts had standing to bring its climate lawsuit.[24] Whatever decision the Justices reached that day would eventually find its way into a new volume, joining the thousands of binding decisions that preceded it.

As the Justices entered the conference room that Friday morning, they all shook hands before taking their seats. On the table before them was a tabletop book stand with the day's agenda, including the petitions for writs of certiorari under consideration for possible review. Directly in front of and apart from the book stand was a horizontal pencil holder, with two sharpened pencils precisely laid out.[25]

On that particular day the Justices denied review in 192 cases and granted review in three, which would be argued that spring. Two of the three were high-profile First Amendment freedom of speech and religion cases in which conservative Justices were looking to reverse lower court rulings. But the most important item on the Justices' agenda was a discussion of the cases argued earlier that week, including *Massachusetts v. EPA*.[26]

Perhaps precisely because the stakes can be so high and the Justices can feel so strongly about the outcome of cases, there is a formulaic, dispassionate ritual to their deliberations about each case, which masks the inevitable suspense and nervousness the Justices feel as they wait to learn the votes of the others in cases about which they care deeply. They may harbor strong suspicions about how others plan to vote, but they never know for sure until the time comes for each Justice to speak. No matter how enormously important or comparatively small-fry a case may be, they follow the same disciplined format for discussing each. The genial banter heard at lunch in their dining room is largely absent, reflecting the seriousness of their tasks at hand.

For each case discussed that morning, the Chief began, as he always does, by summarizing the case and the legal issues before the

Court, and then he briefly stated his view on whether the judgment of the lower court should be reversed or affirmed and the reasons for his view. Justice Stevens spoke up next, briefly describing his vote and reasoning, followed by each Justice in decreasing seniority, ending with Justice Alito. No Justice spoke twice before each Justice had the opportunity to speak. In this forum, even Scalia would have to wait and hold his tongue, out of respect for his colleagues and the Court's traditions.[27]

Massachusetts was the third case they discussed that morning. The Chief did not surprise his colleagues. He began by announcing he was voting to affirm the D.C. Circuit opinion dismissing the case, and he explained why he believed the *Massachusetts* petitioners lacked standing to bring the lawsuit forward. His vote and reasoning were in keeping with the tenor of his questions at oral argument. But they also had deep roots. Since his early days as a Justice Department lawyer during the Reagan and Bush administrations, Roberts had strongly believed in the importance of strict limits on Article III standing to bring lawsuits in federal court.[28] As he later explained in his published opinion in the case, "Global warming may be a 'crisis,' even 'the most pressing environmental problem of our time,'" but "redress of grievances of the sort at issue here 'is the function of Congress and the Chief Executive,' not the federal courts."[29]

Justice Stevens was next. Here too there was no surprise. He voted to reverse the lower court's ruling, largely mirroring the reasons set forth in D.C. Circuit Judge David Tatel's dissent. Massachusetts should, Stevens explained, win all issues. He was satisfied that the petitioners possessed standing, and he thought that greenhouse gases were clearly air pollutants within the plain meaning of the Clean Air Act's definition. He also found unpersuasive the EPA's backup argument that even if greenhouse gases were air pollutants, the agency had offered sufficient reason for not deciding whether greenhouse gas emissions from new motor vehicles could reasonably be anticipated to endanger public health or welfare.

Justice Scalia was next up. He had reason to be tired. He had returned late the night before from a day trip to Harvard Law School, his alma mater, having been invited by Elena Kagan, then the dean, who hosted a tribute to the Justice. The day's celebration was disrupted when a law student challenged Scalia about how he would vote at the conference the next day on the *Massachusetts* case. "It's your responsibility on behalf of my generation, on behalf of your 28 grandchildren," the student admonished the Justice, to address global warming. Scalia characteristically pulled no punches in his response. He scolded the student for inappropriately "arguing a case" that was pending before the Court, and declared simply, "I will be courageous in my vote, and I will do the right thing."[30]

Scalia's definition of "the right thing" was no doubt different from the student's. And he certainly was not one to be cowed by suggestions that his vote should be determined by a student's views on what would be best for Scalia's family rather than Scalia's belief about what the rule of law required. Scalia agreed with the Chief that the petitioners lacked standing and further said that he thought the EPA had the discretion to decline to address the endangerment issue even if greenhouse gases were pollutants. Scalia said he had found nothing in the statutory language that limited the EPA's discretion not to decide the endangerment issue now.

Everyone knew that the next vote, Justice Kennedy's, would likely determine the outcome of the case. As one Justice later put it, it was, "of course, all about Kennedy," because he was "the decisive vote."[31] Unbeknownst to anyone outside the Court itself, the Justices sympathetic to the petitioners had voted the prior June to grant review based on the assumption that Kennedy would be prepared to vote with them. Absent such confidence, they would have opposed review.[32]

Still, despite their optimism, Kennedy's earlier decision to join them in granting review was no guarantee of how he would vote many months later after a full briefing and oral arguments. At most his vote the prior summer had demonstrated he was open to the possibility of ruling against the EPA. But Justices regularly change their minds. Kennedy had disappointed them in the past. And given Justice Kennedy's long-standing concern about excessive federal environmental regulation, his vote was especially susceptible to change.

But as soon as Kennedy began to speak and it became clear that he was joining Stevens in voting to reverse the lower court's decision, liberal Justices like Ginsburg who favored Massachusetts could quietly smile to themselves. If the remaining votes were cast as they had hoped, the case could be a major victory for the liberal wing of the Court in a year in which they had seen few. In explaining his vote, Kennedy focused, as he had in oral argument, on the special status of a state like Massachusetts to bring a lawsuit to protect against loss of its sovereign territory from climate change. (Under the Court's precedent, the Court did not need to find that all the petitioners possessed standing—the lawsuit could be maintained so long as at least one petitioner satisfied the Court's standing requirements.) Kennedy also made it clear that he agreed with Stevens that greenhouse gases were air pollutants and that the EPA had fallen short in its reasons for not yet deciding whether emissions of those gases endangered public health and welfare.

Justice Souter, who said he agreed with Stevens, spoke next, followed by Justice Thomas, who expressed agreement with Scalia and the Chief. Justice Ginsburg then announced that she shared Stevens's view. Her voice was so quiet and at times even halting that the other Justices often had to strain to hear her words. But there was no denying the force of her views.

After the first seven Justices had spoken, the vote was four to reverse the lower court decision and three to affirm. Only two votes remained to be cast: Breyer would speak next and Alito would cast the final vote.

Before the oral argument, the *Massachusetts* petitioners were not expecting Alito to vote in their favor and they were hopeful (but not confident) that Breyer would be supportive. They were correct to harbor some residual concern about Breyer. They knew better than to assume, as too many members of the public mistakenly do, that Justices vote reflexively based on the political ideology of the president who nominated them to the Court.

Both as a Harvard Law professor and as a federal appellate judge and Justice, Breyer had espoused a faith in administrative agency expertise, a wariness of judges second-guessing agency officials, and strong support for the kind of cost–benefit analysis many environmentalists criticized.[33] He had disappointed environmentalists in past cases by arguing that agencies like the EPA should have broad authority to consider compliance costs in setting environmental protection standards. There was a reason the Senate had confirmed Breyer by an 87–9 lopsided vote. He had the strong support of much of the business community.[34]

It was possible that, here too, Breyer would defer to the EPA. He might not be willing to second-guess the agency's reasons for concluding that this was not the right time to address the endangerment determination. To the Carbon Dioxide Warriors, this would be a disappointment, but it would preserve the EPA's authority to address climate change were control of the White House to change parties. But it was also possible, though less likely, that Breyer would be willing to embrace the EPA's view that the Clean Air Act would become unworkable if greenhouse gases were considered "air pollutants" under it. That would be a far worse outcome, because such reasoning would mean that the EPA could never regulate greenhouse gases absent new legislation. It would be a bittersweet loss to have gained Kennedy's vote, typically the hardest to secure, only to lose Breyer's.

In other circumstances, Stevens, Souter, and Ginsburg might have had reason to worry about the reliability of Breyer's vote. But not this time. The oral argument had left little doubt about Breyer's views about the case. His sympathies were clear from his tough questioning of Deputy Solicitor General Greg Garre during the oral argument—and his rescue of Milkey on rebuttal, when Scalia had trapped him into making a problematic concession. The oral argument proved predictive for Breyer. Breyer voted to reverse in favor of the *Massachusetts* petitioners.

Alito spoke last. But no one was likely paying much attention at that point even if they went through the motions of appearing to care. Even before Alito began, all the Justices knew that the petitioners had secured their five-Justice majority. Such is the fate of the junior Justice. Unless the vote is tied four to four when it is the ninth Justice's time to speak—in which case they are the focus of everyone's undivided attention—the others can be easily distracted. Alito voted to affirm the lower court decision, just as his aggressive questioning of Milkey at oral argument had suggested he would do.

Milkey himself had no awareness of the result of the private vote of the Justices at conference. He had been hopeful following the argument, but he did not know on Friday morning that his gamble in forcing his fellow petitioners' hands and seeking Supreme Court review appeared, amazingly, to be paying off.

But any celebration even then would have been premature. It remained an open question whether the resulting majority opinion would be sweeping and historic or narrow and of little precedential importance. Nor was it even a done deal that the five-Justice majority in favor of the petitioners would hold. For the Justices who were hoping both for a broad ruling and to maintain their thin majority, there was still much work to be done within the Justices' chambers to close the deal. The conference vote is critically important, but only when a minimum of five Justices formally "join" a draft opinion does that draft become a final "opinion of the Court" and possess the force of law.

"Justice A" or "Justice B"

Justices can and do change their minds after the conference vote, and far more frequently than people realize. Most of the time it does not change which party wins or loses the case. A vote at conference of 8–1 or 7–2 becomes unanimous on publication. There can be a gravitation toward the majority if one or two Justices conclude that their reasons for disagreement are not sufficiently important to warrant a published dissent.

Far more often, any post-conference shifts in thinking change what the opinion says rather than who wins and loses. Justices regularly change their minds about the relative strength of competing legal arguments once the draft majority and dissenting opinions begin to circulate among the nine chambers. A dissenting opinion may reveal analytical flaws that the written briefs and oral arguments overlooked, and may persuade some Justices that their earlier views were incorrect. Even the Justice assigned the majority opinion, once compelled to fully address all the legal issues in writing and to consider all the practical ramifications of the Court ruling in a particular way, may change his or her mind about the case. Drafting an opinion requires the kind of in-depth analysis that can reveal problems that may have previously escaped the attention of the Justices.

While individual shifts in voting during the opinion drafting process happen all the time, sufficient attrition from the majority to lead to a different outcome happens less frequently. Nevertheless, it remains a distinct possibility when the majority vote at conference is razor-thin. In a closely divided case like *Massachusetts,* a single vote-shift by a Justice who had been in the original majority can change the outcome of the case. Reportedly Chief Justice Roberts shifted his vote in just this way in the case determining the constitutionality of President Obama's Affordable Care Act, voting to uphold the law after originally voting to strike it down.[35] Two decades earlier, Justice Kennedy had changed his mind after conference on whether the

Court should overrule *Roe v. Wade,* depriving Chief Justice Rehnquist of his majority in *Planned Parenthood of Southeastern Pennsylvania v. Casey.*[36] But such complete reversals in high-profile cases are not common. What is far more likely is that one Justice's loss of enthusiasm will prompt the majority opinion to be written more narrowly, leading to a less significant legal precedent.

The ability to persuade others with a well-written opinion is why the identity of the Justice assigned the responsibility for drafting the "opinion of the Court" is pivotal. If the Justice does a good job, the majority stays intact or might even expand. But if the Justice does a bad job, he or she might well "lose" the majority and be forced to write a much more narrowly reasoned opinion in order to keep a majority, or worse yet, be relegated to writing the dissent. As the most senior Justice in the majority at the conference in *Massachusetts,* Justice Stevens had the authority to decide who would draft the majority opinion.

The Court's reliance on seniority in deciding who has the authority to assign opinions is enormously consequential, yet it is not expressed in any written Court rule or document governing Court procedures. Like most everything at the Court pertaining to its decision making, reliance on seniority to determine the authority to assign opinions is based on tradition but nowhere mandated.

Justice Stevens appreciated the stakes when he contemplated to whom he would assign the *Massachusetts* opinion. Although he was junior to the Chief Justice, he had served on the Court far longer than any other Justice: thirty-one years. Stevens had seen firsthand the advantages and pitfalls inherent in opinion assignment. He knew that the identity of the Justice who drafts the opinion of the Court is hugely consequential because that Justice will determine both the substance of the Court's ruling and its precedential impact.[37] There are many possible ways to write an opinion that affirms or reverses a lower court judgment. It can be written narrowly, establishing little precedent that can affect future cases. Or it can be written broadly, potentially establishing sweeping precedent affecting future cases for decades. As one former Supreme Court Justice put it, "If the Chief Justice assigns the writing of the opinion of the Court to Mr. Justice *A,* a statement of profound consequence may emerge. If he assigns it to Mr. Justice *B,* the opinion of the Court may be of limited consequence."[38]

Justice Stevens was no less aware that the more sweeping and consequential a Justice tries to make the draft of an opinion, the greater the risk that the Justice will lose the majority, especially in a closely divided case. Kennedy especially, Stevens knew from past experience, was quite capable of changing his mind after conference. Kennedy has described himself as "more of an agonizer than many of his colleagues" about how to vote. He will try on a position for a bit, see how it feels, and if after a while it doesn't seem right, he will embrace a different position—not because he is indecisive but because the cases are, Kennedy had explained, often "very hard" and worthy of some struggle.[39]

Stevens recalled one particularly bad experience that had served as an important lesson when his good friend Justice William Brennan had made an opinion assignment only eleven days after Kennedy had joined the Court in 1987. Kennedy had voted at conference with Brennan and three other Justices, including Stevens, in favor of the civil rights plaintiffs in a hugely significant employment discrimination case, *Patterson v. McLean Credit Union* (1988). "You try to assign the opinion to someone who can keep the majority," Stevens explained, but "Bill Brennan assigned to himself and Tony [Kennedy] was the swing vote." Brennan then "failed to keep the majority." In a dramatic reversal, Kennedy not only switched sides but "wrote the opinion the other way." Stevens felt that Brennan "was guilty . . . of trying to keep an important case for himself instead of trying to make sure the majority stayed." The result was a devastating loss in a civil rights case that Stevens felt should have gone the other way.[40]

(Congress responded by passing a law in 1991 that limited the effect of the *Patterson* decision.)[41]

Kennedy tended to be a wild card in environmental cases. His vote was hard to predict and, as always, his initial vote at conference was no guarantee of his final vote once a draft opinion was circulated to his chambers for his review and approval. Kennedy had not shied away from voting with the more conservative Justices in limiting the EPA's regulatory authority in the past.[42] His continued support in *Massachusetts* could not be assumed.

To keep Kennedy, Stevens faced a hard choice. He could assign the opinion to himself and try to write it as broadly as possible without losing Kennedy. Or he could assign the opinion to Kennedy, which would maximize the odds of keeping the majority, as Kennedy would feel some institutional obligation to stay with his original vote. For just this reason a senior Justice will sometimes decide to assign the opinion to the most "marginal" Justice in a thin majority.[43] As Stevens explained in discussing an opinion assignment in a different case, "I thought if [Kennedy] wrote it out himself he was more sure to stick to his first vote."[44] But, as Stevens knew, there is a downside to assigning the opinion to the marginal Justice in a case like *Massachusetts*—Kennedy might write a narrow opinion that would not include the kind of language about climate change that Stevens thought was needed.

Stevens's decision was complicated by the fact that the other cases for which he might receive the opinion assignment that month were limited in number and in interest. He was eligible to write the opinion of the Court in only three of the nine recently argued cases[45]—he had dissented in the other six[46]—and *Massachusetts* was the only one of those for which he had opinion assignment authority.

The vote in those two other cases had been unanimous. One had been an antitrust case[47] and the other a patent case,[48] and the Chief Justice—as the senior Justice in the majority—would get to decide whether Stevens would write the opinion in one of them and, if so, which one. Stevens could end up assigned what the Justices and clerks all describe as a "dog"—a boring, uninteresting case.[49]

Stevens decided. He would write *Massachusetts* himself. It was a big, potentially historic case, and he had things he wanted to say about climate change. Only by being in charge of the opinion himself could he do his best to make sure it made the points he thought were the most important. He informed the Chief and told his law clerks, who were always elated whenever their chambers took on an important case. His clerks understood the gravity of their responsibility. Could Stevens both keep the majority and write an opinion that was sweeping enough to change the national conversation on climate change?

18

A Bow-Tied Jedi Master

As soon as Justice Stevens assigned himself the responsibility of drafting the majority opinion, his chambers went to work. The case had taken on heightened significance for Stevens, Souter, Ginsburg, and Breyer, who were increasingly finding themselves in the dissent that year. *Massachusetts* was one of their few remaining possible bright spots. In October Term 2005, when both John Roberts and Sam Alito first joined the Court, the more liberal Justices' worst fears had not been realized. The Court had been unanimous in a remarkably high percentage of cases—45 percent—and split 5–4 in only 13 percent.[1] In the new Term, however, the consequences of Alito's having replaced the far more moderate Justice Sandra Day O'Connor were increasingly evident.

Unanimity had plummeted to 25 percent, and the number of cases decided by the thinnest of margins had jumped from 13 to 33 percent.[2] Justice Kennedy had joined the four more conservative Justices in the vast majority of those cases. In the first high-profile case of the year, *Gonzales v. Carhart,* Kennedy had voted with the conservatives at conference in early November and was now drafting an opinion for the Court, upholding the federal Partial-Birth Abortion

Ban Act, which barred a specific medical procedure for ending late-term pregnancies.[3] In early December, Kennedy had also supplied the vital fifth vote at conference for the Court's ruling that two local school districts in Seattle and Kentucky had unconstitutionally considered race in assigning students to public schools to achieve racial balance in the classroom, a decision that would set off a firestorm of controversy when publicly released in June.[4]

Relations between chambers became increasingly strained as Justice Kennedy was firmly establishing that he was more conservative than Justice O'Connor on most issues, and that he was now the controlling vote. His views prevailed in all twenty-four cases decided by a 5–4 vote that Term. The liberal Justices were so dispirited that Justice Breyer would later lament, in a rare dissent spoken from the bench, "It is not often in the law that so few have so quickly changed so much."[5]

It was against this backdrop that Justice Stevens began drafting the *Massachusetts* opinion, firmly intent on keeping Justice Kennedy on board, but also determined to seize an opportunity that was becoming increasingly rare—to announce a Court ruling on a legal issue that was important to him.

"The Illinois Justice"

Each Justice's chambers consist of a private office for the Justice, an office for administrative assistants, and at least two additional offices for the clerks. Each suite also has its own private bathroom for the Justice—additions that initially were controversial, as Justice Oliver Wendell Holmes worried that eliminating the communal men's room would deprive the Justices of one of the few places where they could informally meet to discuss cases.[6]

The Justices' chambers reflect their distinct personalities, interests, and backgrounds. Stevens's chamber told a story of an extraordinary life. Hanging on the wall immediately behind his desk was a photograph of his parents, Elizabeth and Ernest Stevens.[7] His father was a fabulously wealthy Chicago businessman and hotelier, and Stevens had been raised in the 1920s in one of the nation's largest (three thousand rooms) and most luxurious hotels in the country, known as the Stevens Hotel. As a young boy, John Paul hobnobbed with aviators Charles Lindbergh and Amelia Earhart (and later became a licensed pilot). The hotel lobby included a bronze statue of a young boy modeled after none other than Stevens himself.[8]

An avid Chicago Cubs fan, Stevens went to the 1932 Cubs–Yankees World Series game at Wrigley Field where Babe Ruth famously "called" precisely when and where he would hit a home run—standing in the batter's box and, with great swagger, pointing to the center field stands. The game's scorecard—with all the hits, walks, errors, and outs—hung in his chambers, a ready reminder of his remarkable childhood.[9] In a bookshelf near his desk was a biography of Stevens with the simple title "Illinois Justice."

But his childhood memories were not all good ones, and they contributed to Stevens's appreciation of the harm that could come from abuses of legal authority. When he was fourteen, his father was arrested, convicted, and jailed for embezzlement in a highly publicized trial. The arrest occurred in front of Stevens, and his father's public humiliation left a lasting impression. His uncle, also arrested, committed suicide. Although his father's conviction was later reversed on appeal, the initial arrest and conviction had devastated the family. "A totally unjust conviction," the Justice recounted with great emotion decades later. Stevens knew firsthand that judicial oversight was an important impediment to governmental overreach.[10]

On either side of the photograph of his parents were pictures of his wife, children, and grandchildren, and a signed picture from President Gerald Ford, who had been sharply criticized for nominating Stevens once it became clear that the Justice would not live up to the expectations of conservative Republicans. On matters involving gay rights, affirmative action, environmental protection, private property rights, freedom of speech, and separation of church and state, Stevens not only embraced but increasingly championed legal positions that were antithetical to conservative values. Just over a year before his death in 2006, Ford wrote a letter to be read at a celebration of Stevens's thirtieth year on the Court, making clear his tremendous pride in Stevens's record as a Justice: "I am prepared to allow history's judgment of my term in office to rest (if necessary, exclusively)," he wrote, "on my nomination thirty years ago of John Paul Stevens to the U.S. Supreme Court." Ford's photograph was a powerful reminder of the critical importance of judicial independence.[11]

Stevens had also chosen to prominently display a signed picture of Justice Wiley Rutledge, a champion of civil rights and liberties whom biographers have described as the "conscience of the Court" and the "salt of the earth."[12] Stevens treasured his time clerking for Justice Rutledge, and Rutledge had inscribed the photo "To my friend and former law clerk"—a tradition Stevens had picked up and followed with his own law clerks.[13]

As a clerk, Stevens had assisted Rutledge in the writing of the powerful dissent in *Ahrens v. Clark* in 1948, sharply criticizing the majority ruling for limiting the rights of detained persons to petition a federal court for a writ of habeas corpus to secure their release.[14] Fifty-six years later, now himself as Justice, Stevens wrote the opinion for the Court in *Rasul v. Bush,* ruling in 2004 against the Bush administration and in favor of detainees in Guantánamo Bay, and expressly endorsing much of Rutledge's dissent in *Ahrens.*[15] One year later, he did it again, this time writing the opinion of the Court in *Hamdan v. Rumsfeld,* citing yet another Rutledge dissent (*In Re* Yamashita) striking down as unconstitutional the military commissions President Bush had created to try detainees at Guantánamo Bay.[16]

Stevens loved working with his law clerks. After conference day meetings, he would routinely swing by the larger of the two offices where his four law clerks worked, and sink into a big, worn black leather chair to report on the conference.[17] The clerks were naturally eager to hear what had happened, and whether the Justice thought he would be writing the majority or dissenting opinion in any of the cases. Stevens took pride in his selection of his clerks. He hired students with spectacular academic credentials, most frequently from Harvard and Yale, but also from some midwestern schools, including Michigan, Northwestern, the University of Chicago, and the University of Illinois. He was also prone to hiring students who had done interesting things outside of academics and were, like the Justice himself, more quiet and modest notwithstanding their many accomplishments.

Stevens's four clerks the year that *Massachusetts* was before the Court were Nicholas Bagley, Chad Golder, Jamal Greene, and Lauren Sudeall. All had graduated at or near the top of their law school classes at New York University, Yale, and Harvard. Before going to law school, one had written for *Sports Illustrated* and another had worked for Teach for America. Bagley had previously clerked for the D.C. Circuit's Judge David Tatel.[18]

Supreme Court law clerks can sometimes be weighed down by their own accomplishments, so convinced of their abilities that they fail to exercise the meticulous care the job demands. Boastfulness was not at a premium in Stevens's chambers and, by all accounts, clerking for the Justice was a humbling experience. However bright and accomplished the clerks might be, there was never any question that the Justice had the sharpest legal mind in the chambers. As one former clerk put it, "to watch Stevens ... is to blow your mind" because of the way he "sees argument and counter-argument at the same time"—an extraordinary combination of analytical rigor and insight.[19]

But this analytic brilliance was not the only reason Stevens was revered by his law clerks. Five foot seven inches tall and known for his penchant for bowties, Stevens was unfailingly polite and modest in conversation. If he walked into a clerk's office, he would routinely insist that the clerk remain seated while Stevens leaned or sat on a nearby desk.[20] In contrast to the increasingly combative tone adopted by his colleagues (especially Scalia) during oral argument, Stevens would invariably begin a question with the polite query: "May I ask you a question?"[21]

Stevens was generally considered idiosyncratic in his early years on the Court. His votes defied easy characterization. He seemed far less interested in forging and joining majorities than in writing separate opinions to clarify a point of law for himself. No one ever questioned his brilliance, but his maverick quality undermined his influence on the Court during his first decade as a Justice.[22]

By the time of *Massachusetts v. EPA,* Stevens's role had transformed dramatically. As his seniority rose steadily with the departures of Justices Brennan, Marshall, Byron White, and Harry Blackmun and the arrivals of Justices Souter, Thomas, Ginsburg, and Breyer in the late 1980s and 1990s, Stevens proved increasingly adept at building and keeping majorities closer to his views. He was also willing to compromise when needed so that his opinions would secure and keep the necessary votes for a majority.[23]

Because of his unwillingness to vote regularly with the more conservative Justices appointed to the Court by Presidents Reagan and Bush, Stevens was beloved by liberals and denounced by conservatives. He deflected both plaudits and criticism by joking that he didn't think of himself "as a liberal *at all*" and that he was in fact "pretty darn conservative." What had changed, he always maintained, was not him but the rest of the Court, which had become so much more conservative since he had joined the bench in 1975.

Now approaching his eighty-seventh birthday, Stevens showed no hint of slowing down. The Justice played tennis at least three times a week when he was in Washington, and in Florida, where he lived when the Court was not in session, he supplemented his tennis with golf twice a week and a daily swim in the ocean. He would occasionally read his briefs at the beach and, as a result, sand would sometimes fall out onto the courtroom bench when he opened the brief during an oral argument. He joked that it "made my neighbors a little jealous."[24]

Stevens had forged strong ties over the years with his colleagues on the Court, especially Reagan appointees Justices Anthony Kennedy and Sandra Day O'Connor and George H. W. Bush appointee Justice Souter. The strong mutual respect those Justices held for one another had diluted the ability of Chief Justice Rehnquist and Justice Scalia to forge consistent conservative majorities on a host of issues. Regularly joined by Souter—who, like Stevens, had proved to be more moderate than he had been assumed to be at the time of his nomination—Stevens was frequently able to persuade either Kennedy or O'Connor, and sometimes both, to reject the more conservative positions embraced by Rehnquist, Scalia, and Thomas, after Thomas joined the Court in 1991, replacing Thurgood Marshall.

The upshot was that although the Court tipped decidedly to the right during the 1990s and early 2000s, each year was punctuated by several significant rulings celebrated by political liberals. When that happened, Stevens was not the controlling "swing" vote, but he was the senior Justice responsible for ensuring that the majority initially established in the conference remained intact during the opinion-writing process. This was his job in *Massachusetts*.

"Please Join Me"

As in every case in which he wrote an opinion, Stevens wrote the first draft of *Massachusetts* himself, rather than leaving the task to a law clerk. He had picked up the habit from Justice Rutledge and continued to believe that it was the best way, as he put it, to "be sure you understand the case."[25] Once drafted, the opinion would be turned over to the law clerks, who had the important jobs of making it more readable and ensuring that Stevens didn't "go overboard when [he] shouldn't"—or, as Stevens described it, "to prevent the boss from looking like an idiot."[26]

"Every case is really different," Stevens stressed. In some cases, "you're just trying to give the parties an intelligent answer." But "sometimes you're very interested in making a public statement that you thought was important."[27]

The *Massachusetts* case was very important to him, for he had long felt that the climate issue was a major problem. He was troubled by the fact that so many intelligent Republicans wouldn't even discuss the issue, and he thought his position on the case was "dead right." He didn't have "any doubt," he later confided, about the right outcome. "It was not a hard case," he said. "It was a relatively easy case." He wanted his opinion to be read by the general public and not just by lawyers.[28]

For Stevens, drafting an opinion always began with a careful statement of facts. It was a lesson he had learned soon after joining the federal appellate bench in 1970, when a senior federal judge on the court, Judge John Simpson Hastings, had taken him aside to offer advice on how best to write judicial opinions: "If you do a careful statement of facts," Hastings had said, "the rest of the opinion will write itself." Stevens took this advice to heart, first on the United States Court of Appeals for the Seventh Circuit and later as a Supreme Court Justice.[29]

He decided to begin the *Massachusetts* opinion with a strong statement about the profound importance of the problem of climate change. With that factual premise established, the necessary legal conclusions would follow: how injury from climate change satisfied Article III standing requirements, how greenhouse gases that caused climate change were clearly "air pollutants" subject to the

Clean Air Act, and how important it was for the EPA to better explain its reasons for not regulating such pollutants.

Stevens was under no illusion about the challenge he faced. He had to walk a tightrope while juggling the competing views within his five-Justice majority. "I was always concerned about keeping the majority and not driving people away," he later explained, "and there is that danger."[30] He had seen just that happen a year earlier to Justice Scalia in an important Clean Water Act case, *Rapanos v. United States*, in which Scalia lost Kennedy's vote during the opinion drafting and was relegated to issuing a four-Justice plurality opinion that Kennedy did not join.[31]

In the *Massachusetts* opinion, he needed to make clear at the outset the seriousness of the climate issue, in order to create the factual premise that would support the legal conclusions he wanted the Court's opinion to advance. But he also knew he had to be careful how far he pushed his views on climate, the Article III standing of environmental plaintiffs, and the EPA's responsibility to address the endangerment issue. He could not risk losing Justice Kennedy's fifth vote, and he knew the four dissenters were working hard to change Kennedy's mind.

Scalia would not accept that Kennedy's vote at conference was fixed and unshakable, any more than Stevens would have done if the tables had been turned.[32] The Chief Justice and Scalia had both announced their intent to draft dissenting opinions, with the Chief focusing exclusively on the Article III standing issue and Scalia focusing on whether the EPA had the authority to regulate greenhouse gases and, even more forcefully, whether the agency had discretion to decide not to make an endangerment determination at this time. Both would have Kennedy in their sights and would do their best to dislodge him from the majority. Stevens knew he was crafting an opinion on behalf of an inherently tenuous majority. He was most worried about losing Kennedy to the Chief on standing—and acknowledged this was "the principal issue that divided us"—knowing that his majority opinion on that issue arguably "made new law."[33]

The Chief and Scalia had very different working styles. Scalia was as volatile and uncompromising as the Chief was reserved, polite, and diplomatic. Life in Scalia's chambers has been described by clerks as "like an Italian street fight," in which clerks and the Justice would "conduct a no-holds-barred debate until he decided which way he would vote." As one clerk described the excitement of Scalia's chambers, "he wanted his clerks to test and challenge his own assumptions." "The clerks would argue with one another and with the Justice—loudly, passionately, and utterly without fear that we would offend the Boss."[34]

After working closely with one of his law clerks to revise his initial draft, Stevens circulated his first official draft to the other chambers. The draft began with a strong opening paragraph about the compelling nature of the problem of climate change. The *Massachusetts* petitioners may have decided, for their own strategic purposes, to de-emphasize the climate issue in their briefing and oral argument and to argue that the case called for nothing more than the application of ordinary principles of administrative law. But Stevens wanted to put the climate issue front and center.

Justices respond to opinions when they are circulated with a formal written letter sent through the Court's internal mail system to the opinion's author. The response the author of the draft opinion most wants to see is a letter consisting of little more than the three words: "Please join me."[35] This means that the Justice joins the draft as is and requires no modification. Often the Justice will politely suggest a few possible changes in the opinion, in which case the critical issue is whether their joining the opinion is conditional on the acceptance of the requested changes.[36] The response one most wants *not* to see is that the Justice has decided to await further consideration of a circulating dissenting opinion, or any other response that suggests they are now on the fence.[37]

Three Justices (Souter, Ginsburg, and Breyer) quickly agreed to join Stevens's draft opinion asking for no or few changes, none of which was difficult to make. But Kennedy did not immediately respond with an unqualified "Please join me." That meant Stevens had four votes: one shy of the five necessary for a majority.

The Chief Justice and Justice Scalia responded to Stevens's draft opinion by circulating their own dissenting opinions. Each joined the other's dissent, and Justices Alito and Thomas joined both the Chief's and Scalia's dissent. Kennedy did not join either dissent. There were now eight final recorded votes, equally divided four to four.

Stevens and his law clerk worked doggedly with Kennedy and his chambers throughout the winter and early spring to keep Kennedy in the fold. Again and again, Stevens would send a draft opinion to all the other chambers, wait to hear whether Kennedy had any concerns, only to follow up when he did with a new draft seeking to address those concerns. Stevens's first draft became a second draft, then a third, fourth, fifth, sixth, and seventh as further revisions were made. With every new change, some more significant than others, Stevens sought to do no more than required to secure Kennedy's formal vote.[38]

Kennedy asked Stevens to make significant revisions to two parts of his draft opinion: its discussion of the Article III standing of the *Massachusetts* petitioners and its analysis of the EPA's discretion to decide whether to regulate greenhouse gas emissions from new motor vehicles. Neither set of changes was remotely transformative of the bottom line, but each was essential if Stevens wanted to keep the majority intact.[39]

To satisfy Justice Kennedy on the issue of standing, Stevens modified the draft opinion in several places to put more weight on the special status of a state, like Massachusetts, to protect its sovereign territory from climate change.[40] This was, of course, the point

Kennedy had made at oral argument in questioning Milkey. Kennedy had not only suggested to Milkey that states might have heightened standing in a climate case, he had offered the century-old Supreme Court precedent, *Georgia v. Tennessee Copper,* to support such a ruling.[41]

Milkey had been surprised by Kennedy's reference to a case that neither the petitioners' brief nor that of any of their supporters had once cited. What Milkey had no reason to recall was that almost three years earlier, at the outset of the D.C. Circuit litigation, the petitioners' legal team had in fact considered the applicability of the line of Supreme Court precedent the *Georgia* case represented and concluded that it provided them with only thin support because *Massachusetts,* unlike *Georgia,* involved a case against the federal government. Chief Justice Roberts was in fact relying on this same distinction in his dissent. Happily, not only had Kennedy supplied the *Georgia* precedent they had missed, but because the petitioners had not cited the *Georgia* case, the Solicitor General's Office never had a chance to explain why the case should not be controlling.[42]

Stevens drafted his opinion to make explicit reference to the special status to which states were entitled in Article III standing inquiries, cited *Georgia v. Tennessee Copper,* and otherwise included lengthy quotes from prior opinions written by Kennedy on Article III standing.[43] Stevens, in short, fully embraced Kennedy on the standing issue. The changes he made to keep Kennedy on board were not costless to future climate lawsuits. The opinion now left it less clear whether non-state petitioners like the Natural Resources Defense Council and Sierra Club would have the standing to bring a future climate change case on their own.

To satisfy Kennedy on the matter of the EPA's discretion to postpone a decision on whether greenhouse gas emissions endanger public health and welfare, Stevens also had to make further changes to his draft opinion. The opinion's basic ruling remained the same—the EPA had acted unlawfully in declining to decide the endangerment issue because the specific reasons it had originally offered for its refusal were unreasonable. But as revised, the opinion made it clear that it took no position on whether the EPA could on remand offer legitimate reasons for not deciding the endangerment issue.[44]

Every concession Stevens made to secure Kennedy's vote risked the loss of a vote of a Justice who felt that Stevens's majority opinion was becoming unduly timid. In a formal expression of his disquiet with changes that placed less pressure on the EPA to act on climate change, Justice Souter circulated a draft concurring opinion in mid-March, which Justice Ginsburg joined, that took greater aim at the EPA for its years of stalling on restricting greenhouse gas emissions. Souter's draft underscored how little the United States had historically done to address climate change despite the fact that it was responsible for a disproportionately high percentage of the greenhouse gases that were causing climate change. Souter unfavorably contrasted the minimal nature of the past efforts of the United States to address the climate issue with the far more significant efforts undertaken by Lithuania—a country with far lower greenhouse gas emissions than the United States. Souter's draft included a version of Teddy Roosevelt's admonition: "Do what you can, with what you have, where you are."[45]

Justice Souter did not take the further step of indicating that he would decline to join Stevens's opinion—which would have deprived Stevens of his five-Justice majority—but this separate concurring opinion nevertheless suggested the possibility that Stevens's majority might be fracturing. The longer Kennedy took, the more changes he required, the greater the risk either that Kennedy would have second thoughts or that the accommodations on his behalf might alienate Souter or Ginsburg. That would clearly be bad news—but Souter's draft concurrence also offered Stevens, who was a skilled negotiator, an opening. He could use Souter's move to strengthen his hand with Kennedy, now that he could legitimately claim that there were limits as to how far he could continue to modify the opinion.

Justice Stevens's hard work paid off. In late March, after reviewing the eighth draft, Justice Kennedy finally sent him the letter he had been hoping to receive: Kennedy formally agreed to join the opinion. With Kennedy now officially on board, Stevens had his majority for an "opinion of the Court." The efforts of the Chief Justice and Scalia to dislodge Kennedy had fallen short.

At this point Justice Souter fell back in line. He agreed to withdraw his draft concurring opinion, apparently concluding that it was more important to have a unified majority than to make public his unhappiness with the federal government's position.

Stevens had a clean win. The once-idiosyncratic Justice who during his first decade on the Court had written largely for himself, with little regard to forging majorities, had in his late eighties proved to be the Jedi Master of Justice Brennan's infamous "Rule of Five."

The effort required to keep Justice Kennedy on board was particularly evident in the deliberately imprecise wording and somewhat odd structure of Stevens's analysis of the standing issue. In a clear nod to Kennedy, the Court's opinion stressed the essential role of a state like Massachusetts "in protecting its quasi-sovereign interests" and explained that they were accordingly entitled to "special solicitude in our standing analysis" in a lawsuit brought to vindicate their "procedural right to challenge" the EPA's rejection of Joe Mendelson's petition. But it never clearly explained what role, if any, that "special solicitude" actually played. All the opinion could muster was the vague statement that "with that in mind, it is clear that petitioners' submissions as they pertain to Massachusetts have satisfied the most demanding standards of the adversarial process."

The final pages rejecting the EPA's backup argument were similarly vague and elusive in their reasoning. "EPA has offered no reasoned explanation for its refusal to decide whether greenhouse gases cause or contribute to climate change," Stevens wrote, before adding, "We need not and do not reach the question whether on remand

EPA must make an endangerment finding, or whether policy concerns can inform EPA's action in the event that it makes such a finding." He closed with the somewhat elliptical comment, "We hold only that EPA must ground its reasons for action or inaction in the statute."[46] Sometimes a majority vote is retained by being less rather than more clear.[47]

"Respected Scientists"

There was no lack of clarity in what proved to be the most symbolically important part of Stevens's opinion for the Court—its opening paragraphs. Throughout the course of eight revisions, Justice Kennedy never required Justice Stevens to make significant changes to his opening statement foregrounding the compelling nature of climate change. The very first line of the opinion said it all: "A well-documented rise in global temperatures has coincided with a significant increase in the concentration of carbon dioxide in the atmosphere." Nor did the opinion shy away from staking out a position on what had caused that rise in temperatures: "Respected scientists believe the two trends are related."[48]

It was a virtual call to arms for government agencies to address the threat of climate change. When the Court spoke of "respected" scientists, climate change deniers were not among them. The Supreme Court was willing to say what neither the president nor Congress had so far acknowledged, notwithstanding the mountains of evidence and clear scientific consensus: climate change was real and humankind was responsible.

The opinion further stated that the Court had "little trouble" concluding that "greenhouse gases fit well within the Clean Air Act's capacious definition of 'air pollutant.'"[49] The analysis required only a few paragraphs. "The statute is unambiguous," Stevens wrote. The EPA's authority to regulate greenhouse gas emissions going forward was now clear.[50]

The revisions required to persuade Justice Kennedy to join the majority opinion were not insignificant. But none had disturbed the bottom line. The *Massachusetts* petitioners won big on all three legal issues decided by the Court: their standing to bring the case, the EPA's authority to regulate greenhouse gases, and the unlawfulness of its decision to deflect any determination as to whether their emissions endangered the public health. The Court's ruling would make history.

19

Two Boxes

No one expected the Supreme Court to do anything newsworthy on April 2, 2007. The only item on the Court's official calendar for the day was a public session that morning for routine administrative matters, including the admission of sixty-four attorneys to the Supreme Court Bar. The Justices never disclose ahead of time when they will announce their opinions—let alone *which* opinions will be announced. There was no indication that anything special was planned that morning until a few minutes before ten o'clock, when two ordinary-looking sealed cardboard book boxes appeared in Room 42, part of a suite of offices run by Kathy Arberg, the Supreme Court public information officer. The two boxes, intended for members of the news media, were the first hint that the Court would announce two opinions that morning—one for each box.

The Justices had met in their private conference room the previous Friday. When they entered the room that morning, they found on the table in front of their chairs several sheets of paper placed there by the Chief's chambers. The first sheet listed every Justice by name and the status of their work on opinions, including how many majority opinions they had been assigned to draft but had not yet circulated for review, how many opinions they had circulating, how many they had completed, and how many dissents or concurring opinions that Justice had said they would be writing and their status.[1]

The second set of pages focused in more detail on the circulating opinions. The Justices, in reverse order of seniority, discussed each case for which they had an opinion that was still circulating. They reported on how many Justices had joined the opinion so far, the status of its circulation, and when everyone expected to be done. The Chief's purpose in beginning the conference in this manner, as he always did, was clear: he was putting gentle pressure on each Justice to complete their draft opinions in a timely manner.[2]

When, on Friday, March 30, it was Justice Stevens's turn to speak, he announced that all the votes were in for the *Massachusetts* case. He had five votes, and the majority opinion was ready to be published. Chief Justice Roberts and Justice Scalia each reported that they had four votes for their dissenting opinions and those, too, were ready to be published. After a final scrubbing by the Office of the Supreme Court Reporter of Decisions for any possible formatting errors, all three opinions would be printed in booklet format using the machines located in the basement of the Court. They would then be placed in boxes that would remain sealed until the opinions were announced three days later, when the Court convened on Monday.

The small club of elite reporters who cover the Supreme Court typically arrive before 10 a.m. on any day that the Court is in session. The odds seemed so small that anything important would happen on that Monday that some prominent reporters didn't bother to show up. Bob Barnes of the *Washington Post* was on vacation in Florida.[3] Those who did come sat at their assigned desks, separated by dividers, in the crowded press room located in Arberg's office, ready to spring into action should the Court announce a blockbuster opinion. The reporters are well versed at spinning the seemingly impenetrable intricacies of Court opinions into "breaking news."

Some will peer into Room 42 to see if any sealed boxes are being brought out and will then rush to the courtroom to hear the opinions announced, on the off chance that one might prove to be significant. Only after the Chief Justice has declared in the courtroom that a member of the Court has an opinion to announce will Arberg and her staff unseal the box and distribute the opinion.

When the curtains parted that morning and the Supreme Court Marshal formally announced "The Chief Justice and the Associate Justices of the Supreme Court," there was no reason to suspect that the Justices were about to do anything significant. Justices rarely miss opinion announcements, and only five had bothered to come to the office that day. Justices Stevens, Scalia, Souter, and Breyer were all no-shows.[4]

Much to the chagrin of the Chief, who felt that all members of the Court should be present for all of the Court's public sessions, the four had left town on Friday, like eager college students seeking to "extend" their spring break, taking advantage of the fact that their next conference meeting would not be until April 13 to unilaterally award themselves a full two weeks away from D.C.

At 10 a.m., not one of the Carbon Dioxide Warriors had a clue that the Justices were at that very moment filing into the courtroom to announce their decision. Jim Milkey was at his desk in Boston, flipping through the many environmental cases handled by attorneys in his department—each occupying one of the many distinct piles of paper that littered every conceivable space in his ten-foot-by-fifteen-foot office.[5] Lisa Heinzerling was in her office at Georgetown, getting ready to teach class. Since the oral argument, she had dived back into her teaching and scholarship. Litigation had never been her primary interest, and no doubt she was enjoying being able to write just for herself again and not having to subject her drafts to dozens of lawyers for their review and approval.[6]

Both David Doniger at the Natural Resources Defense Council and David Bookbinder at the Sierra Club had turned their attention to defending state laws seeking to restrict motor vehicle greenhouse gas emissions. Doniger was defending tough new emissions standards adopted by California that both the Bush administration and the auto industry were attacking, and Bookbinder was involved in a trial in Vermont in which the auto industry was challenging Vermont's restriction on motor vehicle emissions, which were modeled on California's. On the morning of April 2, Doniger was in New York running personal errands before joining his ninety-one-year-old mother for Passover dinner that evening.[7] Bookbinder had a late start that morning, and he was driving to downtown D.C. from his home in the Virginia suburbs when the news broke.[8] Joe Mendelson, meanwhile, was with his wife and two daughters, now eleven and nine, on a family spring break beach trip to Anna Maria Island, just south of Tampa, Florida.[9]

Had Deputy Solicitor General Greg Garre been aware that the Court was deciding *Massachusetts,* he might well have viewed the warmer temperature that day as yet another bad omen. The temperature reached 82 degrees—20 degrees warmer than a typical early April day in the nation's capital. It was his practice to sit at the counsel table along with his boss, Solicitor General Clement, on what they called "hand-down days"—days in which the Court was not hearing argument but might release an opinion. For decades now, the solicitor general had made sure there was always a representative of the executive branch—often the solicitor general himself—seated at the counsel table before the Court, ready to accept the Court's mandate. It is a great idea in theory, but a tough one to endure in person if you happen to be sitting there when the Court announces a big case against you.[10]

As soon as the Justices were seated, the Chief Justice announced that Justice Souter had written the opinion in the first case to be announced that morning. In his absence, the Chief would himself read the opinion announcement that Souter had drafted.[11] After doing so, the Chief flatly stated that Justice Kennedy would read the announcement for the second and last case to be decided that day, *Massachusetts v. EPA,* which had been written by Justice Stevens.[12] Even a case as momentous as *Massachusetts* could not keep Stevens from leaving town on the prior Friday for an extra weekend of swimming, tennis, and golf. As soon as Garre heard it was a Stevens opinion, he knew the EPA had lost, though he did not yet know how badly. Like any good professional, he would keep his game face on, showing no emotion as Kennedy spoke.

The statement Kennedy delivered lacked rhetorical flourish. While acknowledging that rises in sea levels had "already begun to swallow Massachusetts' coastal land," it was not until the tenth sentence that the import of the Court's ruling was made clear. Even then, the wording was blandly technical: "We now reverse the judgment of the court of appeals denying the petitions for review," Kennedy declared, reading the words drafted for him by Stevens.[13] Only the actual opinion released by the Court would make plain the significance of the ruling. Years later Garre would joke that he was "convinced to this day that we would have won if it had been snowing" instead of unseasonably warm the morning of argument.[14]

"Good Lawyers Are Making the Difference"

Arberg and her staff listened to the courtroom proceedings from a direct audio feed to a speaker in their office. At the precise instant the Chief announced that Kennedy would be announcing Stevens's opinion in *Massachusetts,* the Court posted a pdf of the majority and dissenting opinions on its website, the box of slip opinions in Room 42 was unsealed, and copies were distributed to members of the news media who were already in the room waiting. Those who chose to hear the opinion announcement in the courtroom itself rushed down to Room 42 soon after, and were immediately handed their own copy by a member of Arberg's staff waiting for them in the hallway.[15] Within minutes the media began to call the parties for comment, and within seconds the number of downloads roared from a few dozen to the thousands.

Because he was counsel of record for the petitioners, within seconds of the announcement in the courtroom the Clerk's Office at the Supreme Court notified Milkey that the case had been decided. He quickly scanned the opinion and spotted at the very end the words he had most wanted to see: "The judgment of the Court of Appeals is reversed." He caught his breath. In early March he had drafted three versions of a press release to be issued by the Massachusetts attorney general in the immediate aftermath of the announcement: the first contemplated a complete victory; the second, a win on whether greenhouse gases were air pollutants but a loss on Article III standing; and the third, a loss on standing and nothing more. Excited and stunned, he reached into one of the piles of papers "filed" on his office floor and pulled out the first version of the press release. In a case he was repeatedly warned they had "next to zero" chance of winning, they had won big.[16]

Bookbinder was still driving to work when his wife called to tell him the news she'd heard on the radio. He immediately turned around and drove home to get the suit he would need for media appearances that day. His wife, a step ahead of him, had one waiting for him, all laid out in the bedroom. Once he was dressed for the cameras, Bookbinder blanketed the media from Sierra Club's downtown office with commentary on the significance of the Court's decision. He stressed that the ruling "gives the EPA authority to regulate greenhouse gases from everything"—not just automobiles.[17]

Doniger was about to take his young son to opening day at Yankee Stadium for an afternoon game when he received a call from an NRDC colleague, who simply said "You won!" He read "little chunks" of the opinion on his Palm Pilot and ended up not making it to the stadium until the sixth inning because he had to field so many media calls. "The Yankees won 9 to 5," Doniger told his son after the game, "and we won 5 to 4."[18]

Heinzerling learned the news only minutes after the Court's announcement. She did one quick media interview with National Public Radio—describing the decision as "a huge deal," and saying, "It is hard to overstate the importance of this"—and then she had to go teach her class of 120 first-year law students.[19] The subject of the class was, fittingly, government processes. She told her students about the Court's ruling, and they burst into applause. At the end of the class, she invited all the students to the Billy Goat Tavern, less than a block from the law school, promising to buy them all a beer. About a hundred took her up on the offer.[20]

By the time Bookbinder reached Joe Mendelson by phone in Florida, CNN already had a crawl at the bottom of his television screen with the "breaking news" that the Supreme Court had decided the greenhouse gas case. Mendelson was elated.[21] He was proud of the catalytic role he had played, beginning with those late nights in 1998 working alone on his petition while his daughters slept. Against all odds, including difficult legal arguments and pushback from powerful environmental groups, they had defeated not just the EPA but, even more, the president of the United States. After all, it was no less than President Bush, prodded by Vice President Dick Cheney, who had dictated the federal government's position on the climate issue that they had just now persuaded the Supreme Court to overturn.

Mendelson quickly began to type up on his laptop a press release on behalf of the International Center for Technology Assessment, which his office back in D.C. issued immediately to major news media outlets. It headlined how the Court had found that the Bush administration had "illegally resisted efforts to regulate global warming pollution" and that the historic ruling "recognizes right to sue because of injuries caused by global warming." "It is clear," Mendelson wrote, "that the legal tools exist right now to address the issue and Congress and EPA should move quickly to ensure that the U.S. starts reducing its greenhouse gas emissions now."[22]

Mendelson didn't receive any media calls in the aftermath of the victory. By contrast, some of the organizations that had strongly opposed his original filing—and then tried to prevent him from filing the first lawsuit to force the EPA to answer the petition—were now playing up their role in the case for the national press. Mendelson figured it was par for the course. Whatever annoyance he felt was overwhelmed by the sheer joy he experienced when he read the Court's ruling.[23]

The *New York Times* ran a front-page, above-the-fold banner headline the next day, declaring "Justices Say E.P.A. Has Power to Act on Harmful Gases," before adding "Agency Can't Avoid Its Authority—Rebuke to Administration."[24] The *Wall Street Journal* was less celebratory. Its front-page headline said "Court Rulings Could Hit Utilities, Auto Makers; White House Strategy toward CO_2 Emissions Is Faulted by Justices."[25] An accompanying editorial mocked the "Jolly Green Justices" for having "granted Al Gore's fondest wish."[26]

Three days after the Court's decision, Earthjustice published a full-page ad in the *Times* celebrating the Court's ruling. Dominating the page was a black-and-white photograph of a newborn baby, eyes shut, tiny fingers nestled against its cheeks, tightly swaddled in a white blanket. Superimposed over the photograph, the ad declared "When the Supreme Court handed down its historic global warming decision on April 2, it wasn't just a bunch of lawyers who won." Below the photograph, text celebrated how the "ruling signals a sea change in environmental protection," before concluding "Good lawyers are making the difference."[27]

Reaction back at EPA headquarters in the Ariel Rios Federal Building on Twelfth Street, Northwest, was split between political appointees and career employees. By then, none of the political appointees who had played a role in the decisions leading up to the case were still there. Jeff Holmstead, the principal architect of the agency's decision to deny Mendelson's petition, was back in private practice at a D.C. law firm. He was shocked when a colleague came to his office the morning of the ruling to tell him they had lost the *Massachusetts* case.[28] The reaction of EPA career lawyers could not have been more different. Two veteran attorneys in the General Counsel's Office, Steve Silverman and John Hannon, may have nominally been on the losing side, but were not so secretly thrilled by the ruling. As Silverman put it, Justice Stevens was his hero. "That decision was one of his greatest achievements, and for me, personally, I could not have been happier."[29]

Green Shades of *Brown*

None of the *Massachusetts* petitioners had ever won an environmental case in the Supreme Court against the federal government after having lost in the lower courts, and for good reason: *No one* had ever done that before. Not once in the Court's two-hundred-plus-year history. But against all odds, notwithstanding the crippling personal acrimony that plagued the legal team in the final weeks of the case, they had won big. The rhetorical sweep of Stevens's opinion far exceeded their grandest expectations.

They experienced shades of how Thurgood Marshall must have felt on the Supreme Court steps in May 1954 when celebrating his extraordinary victory in *Brown v. Board of Education*.[30] Just as Marshall had won a historic ruling outlawing racial segregation in public schools, Mendelson, Milkey, Doniger, Heinzerling, and the others had secured a Supreme Court opinion embracing the urgency of the most important environmental law issue of their time—in a way that allowed for hope that the EPA would now address the problem of climate change before it was too late.

Of course, the petitioners were well aware that what Marshall had accomplished was an order of magnitude greater than what they had just done. By persuading the Court to declare state-sponsored racial segregation in public schools unconstitutional, Marshall had upended sixty years of Supreme Court precedent. A Supreme Court constitutional law ruling has far more staying power than an opinion based on its interpretation of a statute passed by Congress—after all, Congress can always change the law. But environmentalists had won a big case on the most pressing environmental challenge of their lives, against heavy odds, at risk of a devastating loss. They had rolled the dice, and won.

That Justice Stevens's opinion was an enormous victory for anyone concerned about climate change was obvious. Less obvious was how Stevens had carefully taken a page from Chief Justice Earl Warren's playbook in his masterful opinion for the Court in *Brown* itself.

Brown was more than a victory for Thurgood Marshall and for civil rights. It had established the Court's central role in safeguarding individual rights, and accordingly became one of the most celebrated Supreme Court decisions of all time. *Brown*'s celebrity is rivaled only by Chief Justice John Marshall's opinion for the Court in *Marbury v. Madison,* an 1803 case establishing the principle of judicial review—including the power of the Court to strike down as unconstitutional laws passed by Congress and actions taken by the president.[31]

At the time of the Court's ruling, Chief Justice Warren had been on the Court for only six months. As a former governor of California, Warren understood the importance of public acceptance of the Court's rulings, especially on matters as controversial as racial segregation in public schools. The Court has no army to enforce its rulings. It holds no press conferences and mounts no political campaigns to promote its decisions. The Justices rely on the Court's institutional integrity and the persuasiveness of the written word to secure public acceptance of its decisions, including those that may be exceedingly unpopular with some or even a majority of the nation.

Warren wanted people to read and understand the Court's ruling in *Brown.* To that end, his opinion was succinct, clear, and devoid of legal jargon. The text was so short (approximately eighteen hundred words) that it could be reproduced in full in the nation's newspapers, which is exactly what the *New York Times, Washington Post, Chicago Daily Tribune,* and other major newspapers did. The American people could read the ruling and understand why the Court had concluded that "in the field of public education the doctrine of 'separate but equal' has no place."[32]

What went largely unnoticed in the immediate celebration of *Brown* was how Warren had compromised to achieve the unanimity he knew was essential for such a ruling. He had done so by strategically having the Court decide less rather than more. *Brown* established the important new constitutional principle against racial segregation in public schools, overruling decades of precedent. But the Court did so unanimously only by refraining from addressing the crucial question of what relief it would provide to ensure that *Brown*'s promise of desegregation became a reality.

The Court addressed the remedy question one year later, after two more rounds of oral argument. This later ruling, known as *Brown II,* is largely unknown to the public and, unlike *Brown I,* was not accompanied by headlines in the nation's newspapers. Yet *Brown II* determined the extent to which the promise of *Brown I* could be achieved through the courts. In *Brown II* the Court quietly insisted only that desegregation be achieved "with all deliberate speed"—a standard that many public school districts seized upon to maintain racial segregation long after the case was decided.[33]

Justice Stevens had been striving not to achieve unanimity but to maintain his bare majority of five Justices. But like Chief Justice Warren, he had to compromise to achieve his goal by qualifying the formal relief provided to the winning parties. The compromise was not unlike Warren's "all deliberate speed" caveat in *Brown II.* Stevens's opinion acknowledged that although the EPA had unlawfully declined to determine whether greenhouse gas emissions endangered public health and welfare, there might exist lawful grounds open to the EPA, in the aftermath of the Court's ruling, to postpone such an inquiry. To keep Justice Kennedy within the fold, the opinion flatly stated, "We need not and do not reach the question whether on remand the EPA must make an endangerment finding."[34]

This compromise was no surprise to the *Massachusetts* petitioners. They had promoted its availability at oral argument precisely because they understood the likely impossibility of securing a majority without it. Justice Scalia perfectly understood that the *Massachusetts* petitioners could win only with such a compromise, which is why he sought to lure Milkey into rejecting any such thing. It is also why Justice Breyer immediately threw Milkey a lifeline to escape Scalia's rhetorical trap.

"This Gets Amazing—EPA Email"

At first Stevens's necessary compromise looked like it would be of no consequence. Surprisingly, President Bush embraced the Court's decision and promised bold action to address climate change. A few weeks after the ruling, the president held a formal meeting at the White House with the new head of the EPA, Stephen Johnson, and the secretaries of Transportation, Energy, and Agriculture, followed by a high-profile event for the national news media on the White House lawn. Vice President Cheney was notably absent.

Bush made no effort to minimize the Supreme Court's mandate, despite the opinion's actual qualifications. The president stated in no uncertain terms that "the Supreme Court ruled that the EPA must take action under the Clean Air Act regarding greenhouse gases emissions from motor vehicles," and he announced that he was issuing that day a formal executive order directing the EPA and the other agencies "to take the first steps toward regulation that would cut gasoline consumption and greenhouse gases from motor vehicles."[35]

Stephen Johnson followed the president's remarks by characterizing *Massachusetts v. EPA* as a "landmark decision" and by promising to release a draft proposal for greenhouse gas regulation that fall. He stated that his goal was to have a final regulation restricting greenhouse gas emissions from new motor vehicles by the end of 2008, before the next administration would take over. The nation's first-ever regulations limiting greenhouse gas emissions would be a Bush administration legacy—if it acted promptly.[36]

Johnson's background and reputation gave the *Massachusetts* petitioners reason to hope that he would follow through on his promises. When Bush had nominated him two years earlier, environmentalists had praised the choice, some in rapturous terms. Unlike all the other names being bandied about for the position, Johnson had no industry ties. For more than two decades he had served as a career scientist at the EPA. One leading environmentalist described Johnson as a "spectacularly good appointment."[37]

At first he seemed as good as his billing. He did not shy away from making the tough calls needed to enforce requirements against industry and to safeguard public health. In early November he testified before Congress that the agency would have a "proposed regulation out for public notice and comment" by the end of the year. Johnson had created a top-notch team of sixty to seventy EPA employees, career scientists, economists, and lawyers to craft the necessary findings and draft regulations.[38]

By early December the EPA had completed the first major step in its work. The agency sent to the White House the administrator's finding that greenhouse gas emissions from new motor vehicles endangered public health and welfare. This was required before the EPA could impose limits on vehicle emissions. Following a White House review, anticipated to be perfunctory, the EPA would publish its finding in the Federal Register, followed soon after by the publication of its draft regulations to limit greenhouse gas emissions from new motor vehicles. The draft endangerment finding, all ready to go, was close to three hundred pages long.[39]

But the gears then abruptly ground to a halt. The White House sat on the EPA's endangerment finding. There were even whispers that White House officials had deliberately chosen not to open the email containing the proposal to avoid triggering any possible legal obligation to review it. And that is where the EPA's proposal stayed for months: unopened and unreviewed.

By June the White House's stonewalling was an open secret. On June 25 the *New York Times* published a story on it that prompted Comedy Central's "The Daily Show" to broadcast a segment gamely titled "Be Patient: This Gets Amazing—EPA Email." The show characterized the Supreme Court's ruling in *Massachusetts* as having decided that greenhouse gases "are pollutants and not, as the administration had tried to argue, the musky cologne that makes the atmosphere sexy."[40]

In late July the EPA published a document that strangely detailed in many pages the "unequivocal" evidence that climate change and greenhouse gas emissions posed a serious threat to public welfare, but stopped short of making the kind of formal endangerment finding that would allow the EPA to do something about the problem. Johnson explained that he was not making an endangerment finding, notwithstanding the weight of the scientific evidence in its favor, because the Clean Air Act was "an outdated law . . . ill-suited for the task of regulating global greenhouse gases."[41] In other words, the Bush administration had decided in its final few months to address the climate issue just as it had eight years earlier: by punting.

By the time George W. Bush left the White House in January 2009, global atmospheric greenhouse gas concentrations had reached well above the 350 parts per million that climate scientists had characterized as "tolerable."[42] The total amount that had accumulated in the atmosphere over several decades from sources located in the United States far exceeded those added by any other country. Yet, as global concentrations rose from 357 to 387 parts per million from 1993 to 2009—a dramatic increase compared to the previous thirty-five years[43]—consecutive presidential administrations had failed to impose any restrictions on domestic greenhouse gas emissions. The gap between the United States' actual responsibility for climate change and its willingness to act responsibly was widening, to the rising frustration and anger of the rest of the world.

The EPA had been ready to act eight years earlier, when Christine Todd Whitman announced plans to follow through on Bush's campaign promise to regulate greenhouse gas emissions. She was backed up by Treasury Secretary Paul O'Neill, Secretary of State Colin Powell, and National Security Advisor Condoleezza Rice. A lineup of heavyweights. But they proved no match for Vice President Dick Cheney, who, at the behest of the energy industry, persuaded Bush not only to reverse his campaign pledge but to deny the possibility that the EPA had authority to act on climate even if a future president wanted the agency to do so.

Cheney may have effectively kneecapped Christine Todd Whitman in March 2001, but there had been a cost to that end-run around the EPA's career lawyers. Cheney was no lawyer and he had overreached. He confused his *policy* position—that the EPA should not have the authority to regulate greenhouse gases—with the *legal* question of whether it had such authority. The vice president failed to appreciate that the latter question is answered not by the president but by the statutory language of the Clean Air Act, passed by Congress in 1970 and signed into law by President Richard Nixon. Cheney's insistence that the president send a letter to Republican senators expressing the extreme position that greenhouse gases were not air pollutants had dramatically backfired. A more effective strategy would have been simply to ignore Mendelson's petition, as the Clinton administration had done.

The Supreme Court's ruling in *Massachusetts v. EPA* was a historic victory. The Justices had established for the first time the right to bring a lawsuit in federal court for injury caused by climate change. The doors were now open for climate business, paving the way for future climate lawsuits against federal, state, and local governments, as well as industrial polluters. The win on standing promised a new wave of climate-based litigation to restrict greenhouse gas emissions.

In November the nation's voters elected a new president and Congress in the immediate wake of a financial crisis of a magnitude not seen since the Great Depression. It would be up to the next administration to decide whether, in light of the Supreme Court's *Massachusetts* decision, there would be mandatory limitations on greenhouse gas emissions.

20

Making History

The election of Barack Obama on November 4, 2008, was momentous for many reasons. Within hours it became apparent that among them was the likelihood that he would take bold action to address the accelerating threat of climate change. The Carbon Dioxide Warriors who had spent the last decade fighting the White House and the EPA on the climate issue were ecstatic. They would no longer be relegated to being outsiders whose only recourse was filing lawsuits. They could potentially join the new administration as political appointees and would have the kind of access and influence with those in power that business interests had enjoyed in the Bush administration.

The day after the election, Obama named John Podesta as co-chair of his transition team. Podesta was largely unknown outside the nation's capital, but he was well known to environmentalists, especially Doniger, Mendelson, and Bookbinder. Mendelson was now serving as director of policy, climate, and energy for the National Wildlife Federation, one of the nation's largest (with approximately four million members) and most highly regarded environmental organizations. A big step from the two-bedroom townhouse that had served as the "headquarters" for the International Center for Technology Assessment where he had worked after law school, living paycheck to paycheck, and sometimes no paycheck at all.

Podesta enjoyed a reputation as the consummate tactician, highly respected for his integrity and judgment. He had served as Clinton's White House chief of staff during the tumultuous final years of his presidency and had, amazingly, emerged unscathed. Less well known was that Podesta was a member of the board of trustees of the Natural Resources Defense Council Action Fund, the political lobbying organization affiliated with Doniger's NRDC. Long a committed environmentalist, Podesta strongly believed in the urgent need to address climate change. His first job out of law school had been as an environmental attorney at the U.S. Justice Department.[1]

Podesta used his new position and strong connections with environmentalists to put a climate change dream team in place across the administration. On December 15, Obama announced that former Clinton EPA administrator Carol Browner would be joining the White House as assistant to the president for energy and climate change policy, serving as the president's quarterback on climate. She would ride herd over all the agencies and work closely with Congress to ensure that the administration addressed climate change.[2]

Obama nominated Lisa Jackson to head the EPA and Steven Chu to be the new secretary of energy. Both had a deep commitment to ensuring that the administration addressed the threat of climate change. Jackson had formally served as New Jersey's commissioner of environmental protection, where she had championed the regulation of greenhouse gas emissions and joined Massachusetts as one of the state petitioners in *Massachusetts v. EPA*.[3] Chu, a Nobel Prize–winning physicist, opened his Senate confirmation hearing by stressing in his second sentence, "Climate change is a growing and pressing problem." His belief that the federal government urgently needed to take action was clear.[4]

With a nod from Podesta, the president appointed John Holdren, a distinguished climate scientist and Harvard University physics professor, as assistant to the president for science and technology.[5] Peter Orzag, an expert on using economic incentives to limit industrial emissions of greenhouse gases, was chosen head of the president's Office of Management and Budget,[6] and Harvard Law School's Jody Freeman, who had assisted the *Massachusetts* petitioners in the Supreme Court, joined Browner at the White House as the president's counselor for energy and climate.[7] Lisa Heinzerling took a leave from Georgetown to serve first as senior climate policy counsel in Jackson's EPA and later as the agency's associate administrator for the office of policy.[8]

The electoral sweep in 2008 gave the Democrats substantial majorities in both the House and the Senate, with a whopping 78-vote margin in the House and (counting two independents who caucused with the Democrats) a no-less-impressive 18-vote Senate majority, which rose to a filibuster-proof 20 after all the results came in.[9]

Both the Senate majority leader, Harry Reid, and the speaker of the House, Nancy Pelosi, made it clear early on that they were on board with the president's plan to address greenhouse gas emissions. The two most important congressional committees would be the Senate Committee on Environment and Public Works and the House Energy and Commerce Committee. The former would be headed by California senator Barbara Boxer—a climate hawk. But the House was more challenging. The senior Democratic member of the House Energy and Commerce Committee was Michigan's John Dingell, who was ruthlessly effective in getting his way. A bare-knuckles fighter, Dingell was notoriously close to Michigan's auto industry.

With the tacit support of Speaker Pelosi, California's Henry Waxman staged a highly risky coup d'état. He ran against Dingell for the chair position—and he won. Dingell was too beholden to the auto industry to be trusted, especially to promote legislation that would reduce greenhouse gas emissions from motor vehicles. He had not only filed an amicus (friend of the court) brief in the D.C. Circuit supporting the EPA's denial of Mendelson's petition, he had further asked for oral argument time because he considered the EPA's position insufficiently tough against greenhouse gas regulation. So one of the House's most powerful members was unceremoniously ousted in favor of one of its biggest boosters of climate change regulation.[10]

By the time Obama took the oath of office on January 19, 2009, all the pieces were in place. Heinzerling was now in the administration, and Doniger, Mendelson, and Bookbinder were each running the climate policy shops at three of the nation's most influential environmental organizations, fully engaged and ready to get to work.

The Road to Paris

The new president wasted no time in raising the climate issue. "The ways that we use energy," Obama said in his inaugural address, "threaten our planet." He promised to "harness the sun and the winds and the soil to fuel our cars and run our factories."[11] One week later the president spoke directly to the issue from the White House, not mincing words. The "long term threat of climate change, . . . if left unchecked, could result in violent conflict, terrible storms, shrinking coastlines, and irreversible catastrophe." "These are the facts," he underscored, as he announced a bold agenda to limit the nation's greenhouse gas emissions.[12]

Obama understood the global nature of the climate problem and believed that, absent a coordinated effort by all the leading industrialized nations, climate change could not be adequately addressed. No country could solve the problem on its own. It made no difference to global atmospheric concentrations of greenhouse gases where the emissions originated. But the United States had to play a significant part.

The president knew that few other nations would agree to significantly reduce their emissions unless the United States first took major steps to curb its own emissions. Although the United States was no longer the biggest annual source of greenhouse gas emissions—that dubious honor had gone to China since 2005—its share over the past fifty years still overwhelmed those of any other country, and America still emitted far more greenhouse gases per person than any other major global emitter.[13] Obama's challenge was to persuade the rest of the world that the United States was serious about curbing its own emissions—and the clock was ticking. Global atmospheric concentrations were rising steadily toward potentially catastrophic levels.

The White House embraced a two-pronged strategy. Administration officials would begin by mounting an all-out effort to persuade Congress to enact sweeping climate legislation capable of transforming the way that cars, trucks, power plants, and the nation's industries used fossil fuels to produce energy. Such transformation by legislation offered, in theory, the most sensible and least costly way to dramatically reduce greenhouse gas emissions. Only national health care would rival climate change as the administration's top legislative priority.

But even though the Democrats enjoyed healthy majorities in both the House and the Senate, enacting climate legislation would hardly be a cakewalk—especially if they were trying to push a national health care bill at the same time. The same financial crisis and threatened economic collapse that had helped catapult Obama and the Democrats into office would make voters wary of any major governmental measures that were not aimed at addressing their immediate, pressing economic needs. Many powerful business interests opposed to climate legislation would be ready to capture the ears, and votes, of members of Congress from both parties by claiming that addressing climate change would worsen the already fragile economy.

That is why, from the get-go, the Obama White House did not put all of its climate eggs in the congressional basket. It is also why, while aggressively pushing for legislation, the White House almost simultaneously launched its second prong of attack and took full advantage of the EPA's existing statutory authority to regulate greenhouse gases in the absence of new legislation. It converted the Bush administration's loss in *Massachusetts* into an unprecedented lawmaking opportunity.

The backup plan proved prudent. By June 2009 the House succeeded in passing a climate bill—though only by jamming it through by a late night razor-thin margin.[14] Mendelson was in the House gallery the night the House passed the bill. He had heard the House Republicans derisively chant "Bye Bye" to the House Democrats who had voted in favor of its passage—a not-so-subtle indication that Republicans planned to use the vote to defeat them in the next election.[15] The Senate took those threats seriously and never acted on the legislation. The Republican chants proved prescient. In the November 2010 midterm elections, the House flipped to a Republican majority.[16] Any hope of climate legislation was dead.[17]

With the demise of climate legislation, the EPA's effective use of *Massachusetts* soon became the whole ball game. Greenhouse gas concentrations were rising at an alarming rate, and the Obama White House quickly responded to the challenge. It took concrete action to demonstrate the sincerity of its commitment and to persuade other nations that they should take measures to reduce their own emissions.

The president took his first formal action on April 17, 2009, when the EPA announced its proposed finding that greenhouse gas emissions from new motor vehicles endangered public health and welfare.[18] Motor vehicles accounted for about 28 percent of all U.S. greenhouse gas emissions.[19]

Although Lisa Jackson was then only ten weeks into her new job at the EPA, this was an easy finding for her to make: the career scientists, economists, and attorneys who had already done all the legwork during the Bush administration happily resurrected their earlier work and delivered it to the new administration, who no less cheerfully published it. Having taken that important first step, the EPA was off to the races, relying again and again on the *Massachusetts* ruling. Heinzerling, now at the EPA's Air Office, worked closely with the career EPA employees and political appointees to push the agency to be as ambitious as possible. As always, Justice Brennan's "pistol" from Minnesota aggressively promoted her views and was not shy about ruffling feathers.[20]

The EPA published its final endangerment finding in December 2009,[21] followed by the nation's first greenhouse gas emissions limitations applicable to automobiles, trucks, and other new motor vehicles in May of the following year. Those rules were just the first of a series of wide-ranging and demanding new emissions limitations. The May requirements were followed by additional new motor vehicle emissions limitations in October 2012 and October 2016.[22] Doubling the fuel efficiency of new motor vehicles by 2025, the new requirements would prevent six billion metric tons of carbon dioxide from entering the atmosphere and reduce oil consumption by twelve billion barrels.[23]

While Heinzerling was now on the inside, using her experience in the *Massachusetts* litigation to try to make new climate laws, Milkey had moved on. He was proud of his role in the case, but also ready for a new opportunity, feeling somewhat burned out by the challenges of that litigation. He left the Massachusetts Attorney General's Office in 2009, shortly after Obama's inauguration, to become an Associate Justice of the Massachusetts Appeals Court. *Massachusetts* was Milkey's first and last Supreme Court argument. In announcing Milkey's judicial nomination in March 2009, Massachusetts Governor Deval Patrick highlighted Milkey's "prominent role in successfully litigating *Massachusetts v. EPA*, a case that many consider to be the most important environmental law case in history."[24]

The Clean Power Plan

The administration next used *Massachusetts* to take on the power plants, which were then the nation's single biggest source of greenhouse gas emissions, responsible for some 40 percent of U.S. carbon dioxide emissions. In June 2013, not long after winning his second term in office, President Obama personally launched the administration's initiative to regulate power plant emissions. In a speech delivered at Georgetown University, the president referred to the Supreme Court's *Massachusetts* ruling—where the "Court ruled that greenhouse gases are pollutants covered by th[e] Clean Air Act." "It needs to stop," Obama said forcefully, before announcing that he was directing the EPA "to put an end to the limitless dumping of carbon pollution from our power plants, and complete new pollution standards for both new and existing power plants."[25]

The White House knew that taking on the nation's power plants and fossil fuel industry would be orders of magnitude more challenging than the motor vehicle industry had been. Fuel efficiency gains in cars and trucks were hard fought, but the Obama administration had been able to tout them as consistent with national security concerns—allowing the United States to reduce its dependence on foreign oil at a time when many felt that the Iraq war had been a costly gamble to secure more friendly oil supplies in the Middle East.

The auto industry was politically weak at the beginning of the Obama administration. Reeling from the economic crisis, the big three domestic car companies (General Motors, Chrysler, and Ford) had asked for bailouts from the federal government to survive, with both General Motors and Chrysler on the verge of bankruptcy. This gave the White House considerable leverage in negotiating its plan to limit greenhouse gases from new motor vehicles. In announcing the terms of the bailout in late March, Obama made clear his hope that the auto industry would be manufacturing "fuel-efficient cars and trucks."[26] The link was clear. As part of a deal in May, the auto industry agreed not to contest the EPA's first motor vehicle greenhouse gas emissions limitations.[27]

That leverage could not be replicated with the nation's coal-fired power plants. The electric utility industry could be expected to mount a scorched-earth attack on any effort to restrict its greenhouse gas emissions. Along with the coal industry, it claimed that any new EPA regulations would threaten the delivery of reliable, low-cost electricity upon which hundreds of millions of Americans depended. In coal states like West Virginia, roadside billboards began to appear, condemning Obama's "War on Coal."[28] The morning after the House passed climate legislation, the fossil fuel industry had filled the nation's airwaves with ads warning about the cripplingly high electricity rates people would have to pay were the House bill to become law, while lobbyists started working their contact lists and setting up meetings.[29]

Fully aware of these challenges, the EPA's new administrator, Gina McCarthy, undertook a massive effort to develop a plan to regulate power plant emissions that would enable the agency to walk the necessary tightrope. She brought in personnel from the Federal Energy Regulatory Commission who had expertise in the workings of the electricity grids linking power plants to consumers across the nation. EPA regulators wanted to find a way to dramatically cut back greenhouse gas emissions without jeopardizing the reliability of those electricity grids or significantly raising people's power bills. Agency officials worked so closely with environmental organizations that Senate Republicans accused the EPA of improperly colluding with environmentalists in crafting its climate policy. Doniger was singled out for criticism, with a public release of emails between Doniger and EPA officials.[30]

EPA's McCarthy knew she did not have the luxury of time. To have any chance of persuading other nations to join in an international climate accord, the Obama administration would need new restrictions in place by the fall of 2015. Almost every year the United Nations sponsored meetings on climate change—referred to as a "Conference of the Parties on Climate Change" or "COP"—to promote the forging of an international climate accord. The first COP was held in Berlin in 1995, and to date every subsequent COP meeting has fallen short, largely because other countries balked at agreeing to reduce their emissions when the United States had not yet done so. For that exact reason, the Copenhagen meeting (known to insiders as COP 15) in December of Obama's first year in office had bordered on the disastrous, with little progress achieved.[31] Obama's last remaining opportunity would be COP 21, scheduled to be held in Paris in December 2015. That meant he would have to get substantial greenhouse gas reductions in place before then.

The EPA's challenge became exponentially harder in June 2014 when, after publishing its proposed Clean Power Plan for regulating power plant emissions, the agency received more than 4.3 million comments on the proposal. Under settled law, the EPA could not publish its final Clean Power Plan until it had considered all significant comments within those millions, any one of which could serve as a legal basis for challenging the EPA's plan. An enormous job, to say the least. But on October 23, 2015, the EPA published its final Clean Power Plan, having made substantial changes in response to all significant comments. The final plan ran to more than fifteen hundred pages.[32]

The Clean Power Plan was easily the most ambitious set of pollution control rules ever undertaken by the EPA. Under its provisions, greenhouse gas emissions from the nation's power plants would by 2030 be 32 percent below what they had been in 2005. It was a huge reduction. The EPA estimated that benefits of that reduction would exceed the costs by at least $26 billion and as much as $45 billion. It also forecast that consumers would over the long term see their monthly utility rates go down significantly by 2030.[33] The linchpin of the Clean Power Plan's strategy was what it called "generation-shifting": moving some of the current electricity production from existing coal-fired power plants to facilities that could produce that same amount of electricity with lower greenhouse gas emissions. These facilities included both existing power plants using natural gas and nuclear power to produce electricity, and new facilities using renewable energy resources such as wind and solar power.

Obama had made good on the promise he'd made, a week after taking office, that the United States would take serious steps to avoid the catastrophic effects of climate change. With little more than one year left in his presidency, he was now challenging the rest of the world to do the same.

"There Is No Planet B"

By the time negotiating teams arrived in Paris in late November 2015, global atmospheric concentrations of greenhouse gases had topped 400 parts per million for the first time in human history—an increase of 20 parts per million from a decade earlier, and 85 parts per million since 1959.[34] With global emissions dramatically accelerating thanks to higher emissions from China and India, there was a very real possibility that, absent concerted international action, worldwide average temperature would continue to rise, with catastrophic consequences. Floods, famine, drought, sea level rise, wildfires, severe storms, species extinction, and the spread of infectious diseases from climate change were no longer the fatalistic predictions of doomsayers. There was an emerging scientific consensus that such consequences, along with ocean acidification from oceanic absorption of carbon dioxide,[35] would result if climate change was not soon brought under control.

Because of Obama's unilateral actions, the Paris negotiations had a chance of success unlike any of the other prior international meetings. But it was far from a done deal. Todd Stern, the chief U.S. negotiator, acknowledged that an agreement wasn't "in the bag" when he arrived in Paris, and it was no more so a week later. India and China were wary of agreeing to terms that would keep them from building economies that could produce the wealth that the United States and much of Europe had long enjoyed. The world's poorest nations were justifiably angry that, because of their geographic location and lack of resources, they were likely to suffer the greatest and most immediate harm from a problem that had been caused by the wealthier nations.[36]

On Friday, December 11, at the end of the second full week of negotiations, Stern felt he had the makings of a possible deal. That day the UN secretary-general, Ban Ki-moon, put out an ominous warning: there was "no Planet B" if the deal fell apart, no alternative planet on which humankind could survive should climate change ravage our own.[37]

The next day, for the first time ever, virtually every nation in the world pledged to cut or significantly limit the growth of its emissions. The pledge would not solve the problem of climate change. It was not even formally binding, at the insistence of the United States, because a binding international treaty would require U.S. Senate ratification that the Obama administration knew it couldn't get. But the pledges of the Paris Agreement provided a critical first step: up to half of the reductions needed to limit climate change to an increase of 2 degrees Celsius (3.6 degrees Fahrenheit).[38] Absent such significant reductions, scientists forecast that temperatures would increase by as much as 5 degrees Celsius (9 degrees Fahrenheit) by the end of the century, with devastating and cascading environmental and economic consequences.

The United States would not be hit the hardest, but a temperature increase of 5 degrees Celsius would disrupt the nation's critical infrastructure for the delivery of water and electricity, causing dangerous shortages of each. It would substantially reduce the country's agricultural productivity, submerge some of the coastal communities where more than 40 percent of the U.S. population currently lives, and result in major increases in water-borne and insect-borne diseases. Cities like Phoenix would suffer from as many as ninety days with temperatures topping 110 degrees Fahrenheit, a more than tenfold increase from 1976 temperatures.[39]

The only Carbon Dioxide Warrior in Paris was Doniger, who had come there to participate in the negotiations on behalf of the NRDC. Mendelson had participated in the prior United Nations climate meetings in Copenhagen (2009), Cancun (2010), Warsaw (2013), and Lima (2014)—first on behalf of the National Wildlife Federation and then, beginning in 2013, in his capacity as chief climate counsel for the Senate Committee on Environment and Public Works.

Mendelson was not in Paris only because a few months earlier he had left government service to join Solar City, one of the nation's largest installers of solar panels (which later merged with Tesla) as policy director and counsel. He felt good about his work in the Senate, but it was not especially satisfying—his time was mostly spent trying to stop a Republican-controlled Congress from sneaking amendments into bills that would eliminate the EPA's authority to regulate greenhouse gases. In the private sector, he now had the opportunity to work with forward-looking businesses that were promoting innovative technologies designed to eliminate greenhouse gas emissions.[40]

David Bookbinder had left Sierra Club a year before to form a private consulting group with an attorney from ExxonMobil, offering advice to organizations interested in climate policy. He was working to promote a tax on carbon that both environmentalists and industry might support, having concluded that it was better than the kind of cap-and-trade approach favored by the Obama administration and many environmental organizations.[41] And Howard Fox, still with Earthjustice, was playing a major role in defending the Clean Power Plan in court. After two years at the EPA, first in the Air Office and then running its policy shop, Heinzerling was back at Georgetown full time.

On Saturday, December 12, Doniger's official credential hung from his neck. Security had been especially tight in Le Bourget, the Parisian suburb where the meetings were being held only two weeks after deadly terrorist attacks had struck central Paris. He wore a bright red tie with cheerful drawings of suns wearing sunglasses, as he watched the proceedings. Doniger was elated by the prospect of an agreement, but also exhausted. He sometimes felt he was the only person in that vast room who understood the critical role that the *Massachusetts* victory had played in securing the Paris Agreement, and just how unlikely this had seemed a decade earlier.[42]

Doniger had waited a long time for this moment. He had been working on air pollution issues for almost thirty years, beginning with his first job at NRDC after law school. He had been working on an international climate accord for more than a decade, both at NRDC and at the EPA during the Clinton administration. Now he sat poised to tweet the news to the world at the precise moment the gavel came down to signal the existence of an agreement. Instead, he found himself wiping away tears of joy, unable to see the screen in front of him.[43]

Epilogue

Joe Mendelson had planned to stay up late on election night November 2016 to celebrate the returns. Much had been accomplished during the Obama administration to realize the potential created by *Massachusetts* to use the Clean Air Act to address climate change, but there was clearly much critically important work left to be done by an administration that would be led by President Hillary Clinton. The Paris Agreement was a spectacular turning point, yet only an important first step. Years, indeed, decades more work would be needed to reduce greenhouse gas emissions and stabilize atmospheric concentrations of those gases at a safe level. To that end, climate activists who had been tapped to be part of the transition team for what Mendelson (and most of the nation) assumed would be the new Clinton administration already had their flights booked to Washington, D.C., and their apartments rented across the nation's capital, so they could be ready to start their work on day one after the election. There was also a Supreme Court seat to be filled, left vacant by Justice Scalia's death in February, which would solidify a majority on the Court ready to reject industry challenges to the

EPA's greenhouse gas regulations, including those pending against the Clean Power Plan.

There was good reason for Mendelson's excitement and optimism at 8 p.m., just before the election returns started to come in. But, not long after, he instead went to bed early on election night once he saw the crumbling of the Democratic Party's "impenetrable" electoral college firewall in Pennsylvania, Michigan, and Wisconsin, which political pundits had for months declared would ensure Hillary Clinton's victory. "Why watch a disaster unfold?" he thought to himself. He awoke the next morning to an unusual silence that confirmed his worst fears. Donald Trump was the president-elect.[1]

The threat to environmental protection law presented by a Trump administration was without modern historical parallel. George W. Bush had reneged on his campaign pledge to regulate greenhouse gases, but, unlike Trump, he had not campaigned on the preposterous proposition that climate change was a "hoax" "created by and for the Chinese in order to make U.S. manufacturing non-competitive."[2] Nor had Bush embraced a national political party platform that promised a wholesale unraveling of the nation's air and water pollution control laws and the agencies in charge of their administration. And, while not embracing greenhouse gas regulation, the Bush administration had funded important climate research that later provided the scientific basis for much of what the Obama administration had accomplished by regulation.

November 2016 was no mere reprise of the election of Ronald Reagan thirty-six years earlier. Notwithstanding his exaggerated political rhetoric, Reagan had challenged existing environmental law mostly at the margins. And despite his personal popularity, his ability to fulfill any broader reform agenda was persistently curtailed by both Democrats and Republicans in the House and Senate, who supported tough air and water pollution control laws. By contrast, the 2016 election produced Republican majorities in the House and Senate that shared Trump's desire to cut back on federal environmental protection, including restrictions on greenhouse gas emissions.

Trump wasted little time in taking steps to fulfill his deregulatory agenda. He named to head the EPA Oklahoma attorney general Scott Pruitt, who had spent the previous six years launching lawsuits against the Obama EPA, including challenging the lawfulness of the Clean Power Plan. (When Pruitt was forced to resign under a cloud of scandal, Trump replaced him with a former coal lobbyist with the same policy agenda.)[3] Trump also named former Texas governor Rick Perry—who had previously called for the elimination of the Department of Energy—to be his secretary of energy.[4]

Trump himself took an axe to environmental regulation shortly after his inauguration. In March 2017, only two months into his presidency, he signed an executive order designed to unwind every one of Obama's climate policies. He revoked several executive orders that Obama had issued to reduce greenhouse gas emissions in the operations of federal government buildings and the military, and instructed the EPA to consider curtailing and outright repealing the agency's restrictions on greenhouse gas emissions from motor vehicles and power plants.[5]

On August 4, 2017, just over six months into his presidency, Trump formally notified the UN secretary-general of the United States' intent to withdraw from the Paris Agreement. Under the agreement's terms, any such withdrawal would not be effective until November 4, 2020, the day after the next presidential election. Trump also canceled the payment of $2 billion previously promised by the United States to the Green Climate Fund to help developing nations address climate change.[6]

Trump-appointed agency officials were instructed to roll back environmental protection regulations, especially those aimed at addressing climate change. Trump's EPA wasted no time in announcing proposals to cut back on regulations governing greenhouse gas emissions established by the agency under Obama. The proposed rules governing new motor vehicles would freeze fuel efficiency standards at 37 miles per gallon in 2020 and eliminate the prior requirement that vehicles subsequently achieve 50 miles per gallon by 2030. The EPA also announced the elimination of California's long-standing right to impose pollution emissions limitations that are more stringent than the national standards. Twelve other states and the District of Columbia have adopted the tougher California standards, and if the federal government were to succeed in eliminating California's authority to adopt more stringent standards, all those states and the nation's capital would have to follow the more relaxed EPA emissions limitations.[7]

To similar policy ends, the EPA announced its repeal of the Clean Power Plan—the Obama administration's plan to curtail greenhouse gas emissions from coal-fired power plants—and to replace it with a plan that several academic studies concluded would lead to an increase in coal combustion emissions. According to an early estimate by the EPA of its own initial proposed repeal-and-substitute plan, the increase in power plant emissions would result in more than fourteen hundred more premature deaths and fifteen thousand more upper respiratory illnesses each year by 2030.[8] The president's budget request for 2020 further sought to radically reduce the money available to basic scientific climate research, by eliminating the EPA's Global Change Research Office,[9] presumably because such government-sponsored research was undercutting the rationale for the president's deregulatory agenda.

Government scientists in November 2018 had announced sobering news about climate change, underscoring the urgency of the very controls Trump sought to eliminate. The report concluded that annual average temperatures across the country had already increased by 1.8 degrees Fahrenheit since the beginning of the twentieth century, and would increase at least another 2.3 degrees by the middle of the century and by 5.4 to 11 degrees by the end of the century if historical trends continued unabated. The federal government report stated that the evidence was "significant, clear, and compelling . . . that global average temperature is much higher, and is rising more rapidly, than anything modern civilization has experienced, with widespread and growing impacts."[10]

The government report noted that the "earth's climate is now changing faster than at any other point in the history of modern civilization" and laid out in stark detail what the consequences of such temperature increases would be. In dispassionate economic terms, the report described how climate change would result in hundreds of billions of dollars in annual losses to the U.S. economy, including almost two billion lost labor hours each year.[11]

The report described how rising sea levels would threaten $1 trillion in public infrastructure, private industry, and private property along the U.S. coast. Devastating droughts would increase in some parts of the nation while no less devastating downpours and floods would cripple other parts of the country. Agricultural productivity would plummet, power plant efficiency would plunge with rising temperatures (leading to more air pollution), and supplies of safe drinking water would seriously decline.[12]

The many proposed and announced Trump rollbacks were far from being *faits accomplis*. The most significant of these were immediately subject to legal challenge in the courts. Like past efforts to curtail environmental protections, they were no more immune to being struck down than was the EPA's 2003 fateful decision to deny Mendelson's petition. And just as the *Massachusetts* petitioners had chosen to challenge the EPA, many states and the nation's leading environmental organizations quickly joined forces to mount challenges to the Trump administration's actions, including Doniger at the NRDC, Fox at Earthjustice, and attorneys from the Environmental Defense Fund and the attorneys general offices of Massachusetts, New York, and California, all of whom had worked on the *Massachusetts* litigation.

The challengers had far more reason to be optimistic about their likelihood of success than the *Massachusetts* petitioners did back in

2003. The strong scientific and economic bases for greenhouse gas regulation that the EPA developed during the Obama and Bush administrations cannot easily be explained away. As a result, any effort by the EPA to revoke Obama-era regulations is susceptible to being struck down by the courts as arbitrary and capricious, and therefore unlawful. As *Massachusetts* made clear, even presidents have to follow the commands of the law. Early on, the Trump administration lost several major environmental cases in the courts, including a case challenging its efforts to immediately suspend regulations designed to curtail greenhouse gas emissions from cars and trucks and from the production of oil and natural gas. According to NRDC, the administration lost fifty-four of the fifty-nine decided cases the environmental organization brought against Trump's environmental rollbacks in the administration's first three years.[13]

The Obama administration greenhouse gas reduction programs also have a life that is wholly independent of the success or failure of any effort by the Trump administration to undo them. The EPA estimated that the benefits of the Clean Power Plan exceeded its costs by tens of billions of dollars.[14] The emissions reductions envisioned by the Clean Power Plan, even if repealed by the Trump administration, will at some point happen simply because they make good economic sense.

For that same reason, many state and local governments have pledged to follow the blueprint set forth in the Clean Power Plan for how electricity may be generated within their jurisdictions with far lower emissions. They have taken their own steps to increase the use of renewable energy in the production of electricity and to promote energy conservation measures that reduce the need for electricity produced by coal-fired power plants. States discovered that they could simultaneously reduce greenhouse gas emissions and lower pollution that threatens public health without a substantial increase, if any, in the cost of electricity. In 2019 the State of New York enacted a law requiring zero emissions electricity by 2040 and an 85 percent reduction from 1990 levels for emissions from all sources by 2050.[15]

Nor has the rest of the world followed the Trump administration's lead. Just the opposite. Other nations have almost uniformly made clear their intent to honor their Paris commitments, and many pledged to increase their efforts in response to the Trump administration's backtracking. China, in particular, has exploited the void left by the United States by assuming a global leadership role—reducing the growth rate of its own emissions and developing and marketing products that can be sold to businesses and individual consumers in other nations who wish to do the same.[16]

Unlike politicians, business leaders must make decisions based on actual climate science. Executives of publicly owned companies, who face fierce competition from other businesses and have fiduciary obligations to their shareholders, cannot imagine away the scientific consensus that temperature increases, rising sea levels, and extreme weather events will threaten their physical facilities and operations. Nor can they ignore the business opportunities to sell products and services to consumers both in the United States and abroad seeking to reduce their greenhouse gas emissions or otherwise limit the harmful impacts of climate change. Rather than support Trump's efforts to roll back greenhouse gas emission limitations on cars, industry leaders have repeatedly made it clear that they favor tougher controls. Shortly before the White House announced that it would seek to end California's ability to adopt more stringent car emissions standards, four of the nation's largest auto manufacturers declared they would voluntarily meet California's more demanding standards.[17] When the EPA similarly announced its proposals to relax restrictions on the emissions of methane, a highly potent greenhouse gas, many of the biggest oil and gas producers subject to those emissions controls made clear their *opposition* to any such relaxation.[18]

It is too soon to measure the full legacy of the Supreme Court's ruling in *Massachusetts*, which has become required reading in law schools throughout the country. But as the 2016 election soberly reminded those who a year earlier had been uncorking champagne bottles in Paris, the kind of transformative change that *Massachusetts* sought to trigger can begin in a courthouse, but it never ends there.

A major Supreme Court ruling can play a critical role in jumpstarting a too-long-stalled lawmaking process. But transformative change depends on more than winning lawsuits. As environmental public interest lawyers understand, every litigation victory is necessarily provisional. The political and economic interests defeated by environmentalists in court do not merely surrender and disappear. They regroup and file their own lawsuits. And they do not shy away from breaking free of the courtroom by supporting the election of new lawmakers who promote their interests and the passage of new laws to overturn court rulings they disfavor.

That is why environmental progress, over the longer term, requires votes too. Not just the votes of five Justices in one courtroom, but of individual voters across the country and the world. These votes must be in sufficient numbers to elect forward-thinking leaders ready to embrace demanding environmental protection measures, and inspirational enough to mobilize people in favor of the necessary social change.

The climate issue can still be effectively addressed. There is time, despite the drumbeat of increasingly dire predictions and the unnecessary setbacks threatened by shifting U.S. political leadership. The worst consequences can still be avoided. The necessary technology and institutional design changes to reduce future greenhouse gas emissions and to adapt to now-unavoidable climate change either already exist or are within reach. Nor are there insurmountable economic obstacles that prevent those new technologies from being adopted. The greatest economic threat by far to public health and welfare will come not from addressing climate change, but from failing to do so.

But a future that effectively addresses the problem of climate change cannot be merely assumed by simple claims that those who oppose such efforts are "on the wrong side of history." As recent political events in the United States make clear, the ability to make history must be won. It is time for the United States, and the rest of the world, to take the bold, far-reaching steps necessary to curtail global greenhouse gas emissions in order to avert climate change's worst consequences. America has both the opportunity and the responsibility to embrace a leadership role in addressing a problem our own nation played an outsized role in creating. But for that to happen, we will need a fully engaged and politically active citizenry committed to addressing the issue with their votes.

It takes five votes at the Supreme Court to make a majority. But as the *Massachusetts* story makes clear, sometimes one committed person can make all the difference.

Note on Sources

Much of *The Rule of Five* is based on my compilation and review of the extensive documentary record (both public and private) as well as on interviews with those personally involved in the events leading up to the Supreme Court's decision in *Massachusetts v. EPA*. The interviews were conducted with people who were active on opposing sides of the controversy both before and after the start of litigation, including the EPA political appointees and career personnel during the Clinton and Bush administrations, attorneys with private law firms, environmental public interest organizations and state attorneys general offices, and high-ranking government officials and employees of both the D.C. Circuit and the Supreme Court. Many of those interviews were on the record, some were on background, and a few were off the record. Because I respect that law clerks to judges and Justices commit to keep their work strictly confidential, I did not speak with any former judicial law clerks involved in the litigation about their work on the case absent prior express approval from the judge or Justice for whom they had worked.

In addition to personal interviews, many of the sources I contacted generously volunteered to assist me by sharing their own records relating to Mendelson's filing of his petition, the EPA's rejection, and subsequent litigation. Their documents significantly supplemented the available D.C. Circuit and Supreme Court public records, which included all the briefs, exhibits, motions, oral argument transcripts, and other filings by the parties.

The invaluable additional material I received included multiple versions of draft briefs, typed and handwritten notes on meetings, and the extensive email correspondence discussions and debates between the many parties who participated in the relevant decision making on one side of the litigation or the other. I further supplemented the documentary record that was voluntarily provided, with documents I received in response to formal written requests that I filed with the Massachusetts Attorney General's Office and the EPA pursuant to applicable state and federal information disclosure laws. All of these documents proved invaluable in my telling.

Because some of my sources provided personal interviews or email responses to my questions only "on background," the book's endnote references to those interviews or emails preserve their anonymity. That is why the endnotes include general references such as "former D.C. Circuit employee," "*Massachusetts* petitioner," "former EPA employee," "former EPA political appointee," or "Supreme Court employee," without further elaboration. The information is sufficient to indicate the general basis upon which the interviewee possessed firsthand knowledge, without offering the kind of precision that would allow for identification. There were multiple individuals who fit into each category, ranging from career staff employees to Supreme Court Justices. I never spoke on or off the record, or on background, about the case or my research for this book with the Chief Justice of the United States, which I mention only out of an abundance of caution because some readers who know that we were law school classmates and friends might erroneously assume I had.

I have myself litigated for several decades before the Supreme Court in more than forty cases and presented oral argument in fourteen of those cases. I have also informally advised several hundred attorneys with cases before the Court, especially in preparing for oral argument. Much of the general knowledge of the Court's operations and litigation strategy before the Court reflected in this book is based on those years of personal experience. Although, as in many other cases, I provided advice to the *Massachusetts* petitioners at a few stages of the litigation, including oral argument preparation, none of the book's record sources relating to the *Massachusetts* litigation originated from the limited advisory role I served in the case.

Notes

1. Joe

1. Joe Mendelson, email message to Ann Madding, June 11, 2017.
2. Philip Shabecoff, "Global Warming Has Begun, Expert Tells Senate," *New York Times,* June 24, 1988, A1.
3. Albert Gore, *Earth in the Balance: Ecology and the Human Spirit* (Boston: Houghton Mifflin, 1992), 297.
4. Ibid., 14, 269.
5. Scott Bronstein, "Is Clinton Cleaner, Greener than Bush? Arkansas Group Balks as Sierrans Back Governor," *Atlanta Journal-Constitution,* November 2, 1992, A8; Peter Applebome, "Clinton Record in Leading Arkansas: Successes, but Not without Criticism," *New York Times,* December 22, 1991, 30.
6. Bronstein, "Is Clinton Cleaner, Greener than Bush?"
7. Ibid.
8. Michael Isikoff, "Quayle: Gore Has 'Hysterical' Views on Environment," *Washington Post,* August 29, 1992, A5; Richard L. Berke, "Bush Criticizes Clinton as Hard on Car Industry," *New York Times,* August 26, 1992, A18.
9. Dianne Dumanoski, "Gore and the Environment: Some See Double-Edged Sword," *Boston Globe,* July 12, 1992, 10.

10. Adam Nagourney, "For Gore, a Bitter Lesson on the Campaign Trail," *New York Times,* January 27, 2000, A1; R. W. Apple Jr., "The 1992 Campaign: Assessment; Super Tuesday Transforms Campaign," *New York Times,* March 11, 1992, A1.

11. Gore, *Earth in the Balance,* 15.

12. Richard L. Berke, "Gore Walks a Political Tightrope at Kyoto Talks," *New York Times,* December 9, 1997, A3.

13. Roberto Suro, "Pulling Punches on the Environment," *New York Times,* October 5, 1992, A17; Timothy Noah, "Gore Treads Softly as Environmental Point Man, Fearing GOP Efforts to Label Him an Extremist," *Wall Street Journal,* September 16, 1992, A18.

14. Curtis Moore, "How Gore Lost His Balance in Kyoto," *Washington Post,* December 14, 1997, C1; Brad Knickerbocker, "Gore Takes Political Heat on Revised Global Warming Position," *Christian Science Monitor,* December 7, 1997, 5; John M. Broder, "Gore to Join U.S. Team at Global Climate Conference in Japan," *New York Times,* December 2, 1997, A10; "Climate Change: Wirth Departure Comes on the Eve of Talks," *Daily Energy Briefing,* November 20, 1997.

15. John M. Broder, "Gore to Join U.S. Team at Global Climate Conference in Japan," *New York Times,* December 2, 1997, A10.

16. Howard Fineman and Karen Breslau, "Gore Feels the Heat," *Newsweek,* October 27, 1997, 24; Gene Gibbons, "Gore Caught in the Middle on Global Warming Issue," *Reuters,* October 22, 1997.

17. Moore, "How Gore Lost His Balance."

18. S. Res. 98, 105th Cong. (1997).

19. National Oceanic and Atmospheric Administration, "Trends in Atmospheric Carbon Dioxide: Data," https://www.esrl.noaa.gov /gmd/ccgg/trends/data.html.

20. James J. McCarthy, ed., *Climate Change 2001: Impacts, Adaptation, and Vulnerability* (Cambridge: Cambridge University Press, 2001), 75–103.

21. Bill McKibben, "Grading the Environmentalists," *Vogue,* January, 1994, 64.

22. Frances Romero, "Energy Czar: Carol Browner," *Time,* December 2, 2008; David Stout, "7 Utilities Sued by U.S. on Charges of Polluting Air," *New York Times,* November 4, 1999, A1; Oliver Houck,

"TMDLs IV: The Final Frontier," *Environmental Law Reporter* 29, no. 8 (August 1999): 10469.

23. John H. Cushman Jr., "On Clean Air, Environmental Chief Fought Doggedly, and Won," *New York Times,* July 5, 1997, 8; Joby Warrick and John F. Harris, "Clinton Backs EPA's Tougher Clean Air Rules," *Washington Post,* June 26, 1997, A1; Michael D. Lemonick, "Carol Browner: The Queen of Clean Air," *Time,* July 7, 1997, 32.

24. U.S. EPA, "Regulatory Findings on the Emissions of Hazardous Air Pollutants from Electric Utility Steam Generating Units," 65 Fed. Reg. 79825, 79826 (December 20, 2000).

25. Michael Kranish, "Political Landscape Changing on Environment," *Boston Globe,* April 21, 1996, 1; James Gerstenzang, "House Rejects Republican Cuts in EPA Budget," *Los Angeles Times,* November 30, 1995, VYA16.

26. Carol M. Browner, interview by author, April 5, 2018.

27. Former EPA attorney, interview by author, June 21, 2017.

28. Browner, interview.

29. *Departments of Veterans Affairs and Housing and Urban Development, and Independent Agencies, Appropriations for 1999, before the Subcomm. on VA, HUD, and Indep. Agencies of the Comm. on Appropriations,* 105th Cong. 58 (1998) (testimony of Carol M. Browner, Administrator); Browner, interview.

30. Browner, testimony; Patrice Hill, "EPA Not Waiting for Senate to OK Warming Treaty," *Washington Times,* March 7, 1998; Anonymous, "Environmental Provisions in Electricity Restructuring Legislation," internal agency pre-decisional memorandum, 1998.

31. Browner, testimony.

32. Ibid., 200.

33. Former EPA attorney interview, June 21, 2017.

34. Browner, testimony, 201–206.

35. Former EPA attorney, interview by author, March 21, 2018; former EPA attorney interview, June 21, 2017.

36. *Is CO_2 a Pollutant and Does EPA Have the Power to Regulate It?, before the Subcomm. on Nat'l Econ. Growth, Nat. Res. & Regulatory Affairs of the Comm. on Gov't Reform and the Subcomm. on Energy and Env't of the Comm. on Sci.,* 106th Cong. 13 (1999) (testimony of Gary S. Guzy, General Counsel of EPA).

2. Rocking the Boat

1. "IPCC Second Assessment: Climate Change 1995," Intergovernmental Panel on Climate Change (1995), 29–35.
2. Ibid., 29.
3. William J. Clinton, "Remarks on Earth Day," April 21, 1993, https://www.presidency.ucsb.edu/documents/remarks-earth-day; "Warming Up to Global Warming," editorial, *New York Times*, November 6, 1993, 22.
4. Joe Mendelson, email message to Ann Madding, June 11, 2017; *Is CO$_2$ a Pollutant and Does EPA Have the Power to Regulate It?, before the Subcomm. on Nat'l Econ. Growth, Nat. Res. & Regulatory Affairs of the Comm. on Gov't Reform and the Subcomm. on Energy and Env't of the Comm. on Sci.,* 106th Cong. 13 (1999) (testimony of Gary S. Guzy, General Counsel of EPA).
5. Mendelson, email, June 11, 2017.
6. Joe Mendelson, interview by author, March 25, 2015.
7. Mendelson, email, June 11, 2017.
8. Ibid.
9. Ibid.; Brad Knickerbocker, "Environmentalism Extends Its Reach," *Christian Science Monitor,* January 12, 1993.
10. U.S. Environmental Protection Agency, EPA 236-R-98-006, "Inventory of U.S. Greenhouse Gas Emissions and Sinks: 1990–1996," ES3, ES5 (March 1998).
11. Joe Mendelson, email to author, April 10, 2019.
12. Joe Mendelson, interview by author, July 30, 2018.
13. Clean Air Act, 42 U.S.C. § 7521.
14. Clean Air Act, 42 U.S.C. § 7602(g); International Center for Technology Assessment, "Petition for Rulemaking and Collateral Relief Seeking the Regulation of Greenhouse Gas Emissions from New Motor Vehicles under § 202 of the Clean Air Act" (October 20, 1999), 10–11 [hereinafter ICTA, Petition].
15. ICTA, Petition, 7–8, 10–11.
16. Ibid., 7–8.
17. Ibid., 13–26.
18. Ibid., 31.
19. Clean Air Act, 42 U.S.C. §§ 7407(d)(3), 7412(b)(3)&(4).
20. ICTA, Petition, 1.

21. Ibid., 34 n.1.
22. Ibid., 32.
23. Mendelson, email, June 11, 2017.
24. William K. Stevens, "Earth Temperature in 1998 Is Reported at Record High," *New York Times*, October 31, 1992, A32.
25. Guzy, testimony.
26. Mendelson, email, June 11, 2017.
27. Ibid.; Mendelson, interview by author, July 30, 2018.
28. Mendelson, email, June 11, 2017; Joe Mendelson, interview by Ann Madding, April 4, 2018.
29. EPA Mail Office Personnel, interview and office tour by author, April 9, 2018.
30. Ibid.

3. A Turd

1. David Doniger, interview by author, April 26, 2017.
2. Ibid.
3. Former EPA employee, interview by author, June 21, 2017.
4. Doniger, interview.
5. U.S. EPA, "Regulatory Finding on the Emissions of Hazardous Air Pollutants from Electric Utility Steam Generating Units," 65 Fed. Reg. 79,825, 79,826 (December 20, 2000).
6. "Control of Emissions from New and In-Use Highway Vehicles and Engines," 66 Fed. Reg. 7486 (proposed January 23, 2001); Doniger, interview.
7. Richard Lazarus, *The Making of Environmental Law* (Chicago: University of Chicago Press, 2004), 101.
8. Andrew C. Revkin, "Despite Opposition in Party, Bush to Seek Emissions Cuts," *New York Times*, March 10, 2001, A1.
9. James K. Glassman, "Administration in the Balance," *Wall Street Journal*, March 8, 2001, A22.
10. Ron Suskind, *The Price of Loyalty* (New York: Simon and Schuster, 2004), 26.
11. Glassman, "Administration in the Balance"; Paul O'Neill, interview by author, June 4, 2015.
12. Dan Albritton, National Oceanic and Atmospheric Administration, "Global Warming: What We Know & What We Don't," a briefing

for Secretary of State Colin Powell, January 29, 2001 (handwritten notes).

13. Christine Todd Whitman, *It's My Party Too: The Battle for the Heart of the GOP and the Future of America* (New York: Penguin, 2005), 170.

14. Christine Todd Whitman, interview by author, August 14, 2015.

15. Lisa Belkin, "Keeping to the Center Lane," *New York Times,* May 5, 1996, SM6.

16. "The GOP Veepstakes," *CBS News,* June 9, 2000, https://www.cbsnews.com/news/the-gop-veepstakes/.

17. B. Drummond Ayres Jr., "Whitman, in California, Fields the Vice-Presidency Question," *New York Times,* April 30, 1995, 37.

18. David M. Halbfinger, "Two Grades, One Record," *New York Times,* December 26, 2000, A1.

19. New Jersey Department of Environmental Protection, *Sustainability Greenhouse Action Plan* (December 1999), https://rucore.libraries.rutgers.edu/rutgers-lib/36882/PDF/1/.

20. New Jersey Department of Environmental Protection, "Sustainability Initiatives Underway in New Jersey; Corporate and Environmental Leaders Support State's Plan," news release, April 17, 2000, http://www.state.nj.us/dep/newsrel/releases/00_0030.htm.

21. George W. Bush, "Remarks by the President-Elect Announcing the Nomination of Christie Todd Whitman as Administrator of the Environmental Protection Agency," speech, December 22, 2000, Washington, DC, https://www.presidency.ucsb.edu/node/284759.

22. "Christine Todd Whitman Discusses the Bush Administration's Environmental Policy," interview by Robert Novak and Bill Press, *Crossfire,* CNN, February 26, 2001, transcript, http://transcripts.cnn.com/TRANSCRIPTS/0102/26/cf.00.html.

23. Christine Todd Whitman interview, *Frontline,* PBS, January 9, 2007, transcript, https://www.pbs.org/wgbh/pages/frontline/hotpolitics/interviews/whitman.html.

24. Christine Todd Whitman, "Remarks of Governor Christine Todd Whitman, Administrator, United States Environmental Protection Agency, at the G8 Environmental Ministerial Meeting Working Session on Climate Change," speech, March 3, 2001, Trieste, Italy,

https://archive.epa.gov/epapages/newsroom_archive/speeches/ef9a581
27adb3b4b8525701a0052e348.html.

25. President Bush to Senators Hagel, Helms, Craig, and Roberts,
    March 13, 2001, correspondence, https://georgewbush-whitehouse
    .archives.gov/news/releases/2001/03/20010314.html.

26. Whitman, *It's My Party Too,* 175.

27. O'Neill, interview.

28. Suskind, *The Price of Loyalty,* 122.

29. Barton Gellman, *Angler: The Cheney Vice Presidency* (New York:
    Penguin Press, 2009).

30. National Energy Policy Development Group, *Reliable, Affordable, and
    Environmentally Sound Energy for America's Future* (Washington,
    DC: U.S. Government Printing Office, 2001).

31. Ibid.

32. Haley Barbour to Vice President Dick Cheney, memorandum,
    March 1, 2001.

33. Suskind, *The Price of Loyalty,* 118–119.

34. Ibid., 120.

35. Ibid., 120–122; Whitman interview, *Frontline.*

36. Gellman, *Angler,* 89–90.

37. Suskind, *The Price of Loyalty,* 123.

38. Whitman interview, *Frontline.*

39. Suskind, *The Price of Loyalty,* 125.

40. Bush to Hagel et al., correspondence.

41. Marbury v. Madison, 5 U.S. 137, 177 (1803).

42. Barbour to Cheney, memorandum.

4. An Agency Misstep

1. Marianne Horinko, interview by author, May 10, 2017.

2. Jonathan Z. Cannon, memorandum to Carol M. Browner, EPA
   Administrator, April 10, 1998; *EPA Testimony Statements, before a
   Joint Hearing of the H. Subcomm. on Nat'l Econ. Growth, Nat.
   Res., and Regulatory Affairs of the Comm. on Gov't Reform, and the
   H. Subcomm. on Energy and Env't, Comm. on Sci.,* 106th Cong.
   (October 6, 1999) (testimony of Gary S. Guzy, General Counsel
   of EPA).

3. Chevron U.S.A., Inc. v. Nat. Res. Def. Council, Inc., 467 U.S. 837 (1984).

4. Former EPA employee, interview by author, March 30, 2015.

5. Ibid.

6. George Bush, "Appointment of Jeffrey R. Holmstead as an Associate Counsel to the President," October 5, 1990, https://www.presidency.ucsb.edu/node/264945; David L. Hancock, "Utahn Who Wanted to Play Basketball Goes to Court for Bush," *Deseret News,* August 30, 1991, https://www.deseretnews.com/article/180653/utahn-who-wanted-to-play-basketball-goes-to-court-for-bush.html; Jeffrey Holmstead, interview by author, August 17, 2018.

7. Bush, "Appointment of Jeffrey R. Holmstead."

8. The White House, "Jeffrey R. Holmstead," https://georgewbush-whitehouse.archives.gov/government/holmstead-bio.html.

9. Holmstead, interview.

10. *Nominations of the 107th Congress, Hearings before the S. Comm. on Env't and Public Works,* 107th Cong. 11 (May 17, 2001) (statement of Jeffrey Holmstead, Nominee for Assistant Administrator, Office of Air and Radiation, EPA).

11. Holmstead, interview.

12. Ibid.

13. Complaint for Declaratory Relief and Writ of Mandamus or Other Order, at 6, *International Ctr. for Tech. Assessment v. Whitman,* No. 02-CV-2376 (D.D.C. filed Dec. 5, 2002).

14. 5 U.S.C. §§ 555(b), 706(1).

15. Joe Mendelson, interview by author, March 25, 2015; Joe Mendelson, email message to author, July 3, 2019.

16. Joe Mendelson, interview by author, March 25, 2015.

17. David Bookbinder, interview by author, July 17, 2018.

18. Ibid.

19. Mendelson, interview, March 25, 2015.

20. Complaint, *International Ctr. for Tech. Assessment v. Whitman.*

21. Ibid., 1.

22. Mendelson, interview, March 25, 2015.

23. Robert E. Fabricant to Marianne L. Horinko, memorandum, "EPA's Authority to Impose Mandatory Controls to Address Global Climate Change under the Clean Air Act," August 28, 2003.

24. "Control of Emissions from New Highway Vehicles and Engines," 68 Fed. Reg. 52,922 (September 8, 2003).

25. David Doniger, interview by author, April 26, 2017.
26. President George W. Bush to Members of the Senate on the Kyoto Protocol on Climate Change, March 13, 2001.
27. Holmstead, interview.
28. Fabricant to Horinko, memorandum, 1 (emphasis added).
29. Ibid.
30. "Standards of Performance for New Stationary Sources and Guidelines for Control of Existing Sources: Municipal Solid Waste Landfills," 61 Fed. Reg. 9905, 9905 (Mar. 12, 1996).
31. Fabricant to Horinko, memorandum, 10.
32. 42 U.S.C. § 7602(g).
33. David Stout, "E.P.A. Chief Whitman Resigns," *New York Times,* May 21, 2003.
34. "Control of Emissions from New Highway Vehicles and Engines" 68 Fed. Reg. at 52933.
35. Holmstead, interview.
36. "Control of Emissions from New Highway Vehicles and Engines," 68 Fed. Reg. at 52929.
37. Ibid., 52925.
38. Mendelson, interview, March 25, 2015.
39. "Control of Emissions from New Highway Vehicles and Engines," 68 Fed. Reg. at 52929.
40. Ibid., 52929–52931.
41. Ibid.
42. Jeffrey Holmstead, email message to Prudence Goforth, Bill Wehrum, and Nancy Ketcham-Colwill, August 28, 2003.
43. Former EPA employee, interview by author, March 30, 2015; former EPA employee, interview by author, July 22, 2015.
44. "EPA General Counsel Resigns," *Environment New Service,* August 15, 2003; Robert E. Fabricant, General Counsel, Volcano Partners New Jersey LLC, letter to Ms. Amy Legare, Chair, National Remedy Review Board, U.S. EPA, November 14, 2012.
45. Nancy Ketcham-Colwill, email message to Lisa Jaeger, Lisa Friedman, John Hannon, and Patricia Embrey, August 28, 2003.
46. Nancy Ketcham-Colwill, email message to other EPA employees, August 28, 2003.

5. The Carbon Dioxide Warriors

1. Joe Mendelson, interview by author, March 25, 2015.
2. David Bookbinder, email message to Peter Van Tuyn and various *Massachusetts* petitioners, November 6, 2003.
3. Joe Mendelson, "9/3 Meeting Notes" (unpublished notes, September 3, 2003), paper file; Joe Mendelson to Global Warming Litigation Group, memorandum, "Re: Coordination of the Legal Challenge on EPA's Greenhouse Gas Petition Ruling," September 8, 2003; Joe Mendelson, "9/17 Conference Call Notes" (unpublished notes, September 17, 2003), paper file; Joe Mendelson, "Agenda of Climate Change Project Meeting" (unpublished notes, September 23, 2003), pdf file; Joe Mendelson, "10/1 CO_2 Conference Call Notes" (unpublished notes, October 1, 2003), paper file; Joe Mendelson, "Global Warming Conference Call Notes" (unpublished notes, October 8, 2003), paper file; Joe Mendelson, "10/10 Conference Call Notes" (unpublished notes, October 10, 2003), paper file; Joe Mendelson, "10/16 CO_2 Conference Call Notes" (unpublished notes, October 16, 2003), paper file.
4. David Doniger, interview by author, April 26, 2017.
5. Ibid.
6. Jeremy P. Jacobs, "Lisa Heinzerling Won't Back Down," *E&E News*, May 27, 2014, https://www.eenews.net/stories/1060000220.
7. Ibid.
8. Former EPA political appointee, interview by author, March 21, 2018.
9. Jim Milkey, interview by author, February 18, 2015.
10. Ibid.
11. Ibid.
12. Colin Adamson and Patrick Sawer, "The South Gets a 90MPH Battering," *Evening Standard,* October 30, 2000, 2–3.
13. Pardeep Pall et al., "Anthropogenic Greenhouse Gas Contribution to Flood Risk in England and Wales in Autumn 2000," *Nature* 470 (February 17, 2011): 382–385; Sid Perkins, "Rising Temperatures Bringing Bigger Floods," *Science,* February 16, 2011.
14. Rob Edwards, "US Greed Leaves the World Not Waving but Drowning," *Sunday Herald,* November 26, 2000, 3; Gary Ralston, "Grave New World," *Daily Record,* November 15, 2000.

15. Michael Settle, "Bush's Decision to Rat on the Kyoto Treaty Is Grim News . . . ," *Herald,* March 30, 2001, 3.
16. John Ingham, "Heat's on Bush over Warming," *Daily Express,* June 15, 2001, 15; Ian Black, "Street Clashes Greet the 'Toxic Texan,'" *Guardian,* June 15, 2001.
17. Milkey, interview.
18. David Bookbinder, interview by author, July 17, 2018.
19. Ibid.

6. Weakness in Numbers

1. Center for Biological Diversity, "States, Environmental Groups Challenge Bush on Global Warming," news release, October 23, 2003, https://www.biologicaldiversity.org/news/press_releases/warming10 -23-03.htm.
2. David Bookbinder, interview by author, July 17, 2018.
3. Joe Mendelson, "Conference Call on CO_2 Petition Case Notes" (unpublished notes, January 28, 2004), paper file; Jim Milkey, email message to Joe Mendelson and various *Massachusetts* petitioners, February 3, 2004.
4. 42 U.S.C. § 7602(g).
5. Chevron U.S.A., Inc. v. Natural Resources Defense Council, Inc., 467 U.S. 837 (1984).
6. Food and Drug Administration v. Brown & Williamson Tobacco Corp., 529 U.S. 120 (2000).
7. Jim Milkey, email message to Joe Mendelson, March 15, 2004.
8. David Doniger, email message to Jim Milkey, March 15, 2004; Joe Mendelson, email message to Aaron Livingston, March 15, 2004; Marc Melnick, email message to Bill Pardee and various *Massachusetts* petitioners, March 19, 2004; David Bookbinder, email message to Jim Milkey, March 23, 2004.
9. Marc Melnick, email message to Jim Milkey, May 12, 2004.
10. Jim Milkey, email message to Marc Melnick and various *Massachusetts* petitioners, May 24, 2004.
11. David Doniger, email message to Jim Milkey and various *Massachusetts* petitioners, May 27, 2004.
12. Bill Pardee, email message to David Doniger, May 27, 2004.

13. David Bookbinder, email message to Howard Fox and various *Massachusetts* petitioners, June 9, 2004.

14. Jim Milkey, email message to Howard Fox and David Doniger, June 14, 2004.

15. David Doniger, interview by author, April 26, 2017.

16. Jim Milkey, interview by author, February 18, 2015.

17. Joe Mendelson, "1/9 Meeting Notes" (unpublished notes, January 9, 2003), paper file.

18. Doniger, interview.

19. David Bookbinder, email message to Jim Milkey and various *Massachusetts* petitioners, January 10, 2004; Marc Melnick, email message to Bill Pardee and various *Massachusetts* petitioners, January 12, 2004; Joe Mendelson, email message to Marc Melnick, January 12, 2004; Jim Milkey, email message to Joe Mendelson and various *Massachusetts* petitioners, January 14, 2004; David Doniger, email message to various *Massachusetts* petitioners, January 14, 2004; Marc Melnick, email message to Bill Pardee and various *Massachusetts* petitioners, January 15 2004.

20. Howard Fox, email message to David Bookbinder and various *Massachusetts* petitioners, March 16, 2004.

21. Nicholas Stern, email message to Bill Pardee and various *Massachusetts* petitioners, March 18, 2004; Jim Milkey, email message to Howard Fox, March 22, 2004; Marc Melnick, email message to Bill Pardee and various *Massachusetts* petitioners, April 2, 2004.

22. Nicholas Stern, email message, March 18, 2004.

23. Mark Melnick, email message to Bill Pardee and various *Massachusetts* petitioners, April 2, 2004.

24. David Doniger, email message to Nicholas Stern and various *Massachusetts* petitioners, March 18, 2004.

25. Nicholas Stern, email message, March 18, 2004.

26. Jim Milkey, email message to Marc Melnick, May 12, 2004.

27. Jim Milkey, email message to Howard Fox, May 19, 2004.

28. *Massachusetts* petitioner counsel, email message to author, December 7, 2018; *Massachusetts* petitioner counsel, interview by author, January 16, 2019.

29. Former Federal Court of Appeals Judge, The Honorable Patricia Wald, "Notes on *Massachusetts* Petitioners' Draft D.C. Circuit Brief" (unpublished notes, May–June 2004), paper file.

7. Three Judges

1. 415 F.3d 44 (D.C. Cir. 2005).
2. Federal Judicial Center, "Biographical Directory of Article III Federal Judges, 1789–Present," https://www.fjc.gov/history/judges.
3. Christopher A. Cotropia, "Determining Uniformity within the Federal Circuit by Measuring Dissent and En Banc Review," *Loyola of Los Angeles Law Review* 43, no. 3 (2010): 815.
4. Jeffrey Rosen, "The Next Court," *New York Times Magazine,* October 22, 2000, 74.
5. Ibid.; Bernard Weinraub, "Reagan Says He'll Use Vacancies to Discourage Judicial Activism," *New York Times,* October 22, 1985, A1.
6. Canton, North Carolina, "About: Facts and Figures," http://www.cantonnc.com/facts-and-figures/; Ruth Marcus and Sharon LaFraniere, "North Carolina Judge Is Seen as Choice for Appellate Vacancy Here," *Washington Post,* September 27, 1986, A15; Peter Applebome, "Judge in Whitewater Dispute Rewards Faith of His Patron," *New York Times,* August 17, 1994, A1.
7. Judge David S. Tatel, "Portrait Presentation Ceremony: Judge David B. Sentelle" (remarks, April 5, 2013), 18–19, http://dcchs.org/usca/JudgeSentellePortraitTranscript.pdf.
8. David Johnston, "Appointment in Whitewater Turns into a Partisan Battle," *New York Times,* August 13, 1994, A1.
9. Arthur Raymond Randolph Jr., "Oral History Project of the Historical Society of the District of Columbia Circuit," interview by E. Barrett Prettyman Jr. (March 15, April 19, and May 17, 2002, and March 1, 2004): 3–6.
10. Ibid., 11–12.
11. Ibid., 9–11.
12. Ibid., 14, 78–89.
13. Ibid., 32, 60.
14. Linda Greenhouse, "Bork's Nomination Is Rejected, 58–42; Reagan 'Saddened,'" *New York Times,* October 24, 1987, A1.
15. Federal Judicial Center, "Tatel, David S.," https://www.fjc.gov/history/judges/tatel-david-s.
16. Ibid.; Barbara Slavin, "A Judge of Character: Although He's Blind, David Tatel Skis, Runs and Climbs Mountains. By Summer's End, He May Be a Top Jurist Too," *Los Angeles Times,* July 28, 1994, E1.

17. United States Court of Appeals District of Columbia Circuit, "David S. Tatel," https://www.cadc.uscourts.gov/internet/home.nsf/Content/VL+-+Judges+-+DST; Edie Tatel and David Tatel, "Foundation Fighting Blindness Hope & Spirit Award" (biographies, Washington, DC, May 2, 2017).
18. Slavin, "A Judge of Character"; Rosen, "The Next Court."
19. Tatel, "Portrait Presentation."
20. Rosen, "The Next Court."
21. Ibid.

8. Completely Confused

1. David Doniger, interview by author, April 26, 2017; Transcript of Oral Argument, at 3, Massachusetts v. EPA, 415 F.3d 50 (D.C. Cir. 2005) (No. 03-1361).
2. Transcript of Oral Argument, 4.
3. Ibid., 8.
4. Former D.C. Circuit employee, interview by author, March 2, 2018.
5. Final Brief for the Petitioners in Consolidated Cases, *Massachusetts*, 415 F.3d 50 (D.C. Cir. 2005), No. 03-1361.
6. Transcript of Oral Argument, 12–25.
7. United States v. Krizek, 192 F.3d 1024 (D.C. Cir. 1999).
8. 467 U.S. 837 (1984).
9. Former EPA employee, interview by author, March 30, 2015.
10. Former EPA political appointee, interview by author, June 1, 2017.
11. 529 U.S. 120 (2000).
12. Brief of Respondent Environmental Protection Agency, *Massachusetts*, 415 F.3d 50 (D.C. Cir. 2005), No. 03-1361.
13. Jeffrey Holmstead, interview by author, August 17, 2018.
14. Former EPA political appointee, interview, June 1, 2017.
15. Transcript of Oral Argument, 38.
16. Ibid., 39.
17. Ibid., 44.
18. Ibid., 47.
19. Ibid.
20. Ibid., 48.

9. A Clarion Dissent

1. Jeffrey Bossert Clark, "Climate Change Litigation, Presentation to the Interagency Working Group on Climate Change Science and Technology, May 18, 2005" (power point slides).
2. Ibid.
3. Lujan v. Defenders of Wildlife, 504 U.S. 555, 560–561 (1992).
4. Richard Lazarus, "Restoring What's Environmental about Environmental Law in the Supreme Court," *University of California at Los Angeles Law Review* 47, no. 3 (2000): 703, 749–752.
5. *Massachusetts,* 415 F.3d at 59 (Sentelle, J., dissenting in part and concurring in the judgment).
6. Brief of Respondent Environmental Protection Agency, *Massachusetts,* 415 F.3d 50, No. 03-1361, 15–18.
7. Former EPA employee, interview by author, July 22, 2015.
8. Transcript of Oral Argument, at 3, Massachusetts v. EPA, 415 F.3d 50 (D.C. Cir. 2005) (No. 03-1361), 18–20.
9. Massachusetts v. EPA, 415 F.3d 44, 53 (D.C. Cir. 2005) (opinion of Randolph, J.).
10. Judge David S. Tatel, interview by author, January 29, 2018.
11. Christopher A. Cotropia, "Determining Uniformity within the Federal Circuit by Measuring Dissent and En Banc Review," *Loyola of Los Angeles Law Review* 43, no. 3 (2010): 815.
12. United States v. Phillip Morris USA, Inc., 396 F.2d 1190 (D.C. Cir. 2005) (Tatel, J., dissenting).
13. Judge Tatel, interview.
14. Former D.C. Circuit employee, interview by author, March 2, 2018.
15. Ibid.
16. Massachusetts v. EPA, 433 F.3d 66, 67 (D.C. Cir. 2005) (mem.) (Tatel, J., dissenting from denial of reh'g en banc).
17. Former D.C. Circuit employee, interview by author, March 2, 2018.
18. Ibid.
19. Ibid.
20. Former D.C. Circuit employee, interview by author, March 6, 2018.
21. Former D.C. Circuit employee, interview, March 2, 2018.
22. Sharon Walsh, "D.C. Lawyer Nominated to U.S. Appeals Court," *Washington Post,* June 21, 1994, B1; Jeffrey Rosen, "The Next Court," *New York Times Magazine,* October 22, 2000, 74; Barbara

Slavin, "A Judge of Character: Although He's Blind, David Tatel Skis, Runs and Climbs Mountains. By Summer's End, He May Be a Top Jurist Too," *Los Angeles Times*, July 28, 1994, E1.

23. Rosen, "The Next Court."
24. Former D.C. Circuit employee, interview, March 2, 2018.
25. Rosen, "The Next Court."
26. *Massachusetts*, 415 F.3d, at 50.
27. Ibid., 53–59 (opinion of Randolph, J.).
28. Ibid., 59–61 (Sentelle, J., dissenting in part and concurring in the judgment).
29. Ibid., 61, 73–74, 77 (Tatel, J., dissenting).
30. Joe Mendelson, "7/20 Meeting Notes" (unpublished notes, July 20, 2005), paper file.

10. Hail Mary Pass

1. Frances Beinecke correspondence to Attorney General Tom Reilly, August 25, 2005 (contemporaneous handwritten annotations by Jim Milkey of telephone call).
2. Ibid.
3. Ibid.
4. Jim Milkey, email message to Kimberly Massicotte and various *Massachusetts* petitioners, July 21, 2005.
5. David Doniger, memorandum to Jim Milkey, undated.
6. Massachusetts v. EPA, 433 F.3d 66, 67 (D.C. Cir. 2005) (Tatel, J., dissenting).
7. "The Statistics," *Harvard Law Review* 131, no. 1 (2017): 410.
8. "New Anthony M. Kennedy Chair at McGeorge School of Law in Sacramento," SCOTUSblog, October 15, 2018, video, 15:15, www .scotusblog.com/media/new-anthony-m-kennedy-chair-at-mcgeorge -school-of-law-in-sacramento.
9. Jim Milkey, email message to Jerry Reid and various *Massachusetts* petitioners, December 23, 2005.
10. David Doniger, interview by author, April 26, 2017.
11. *Massachusetts* petitioners, "Draft Petition for Writ of Certiorari" (unpublished draft, January 26, 2006), Microsoft Word file.
12. "Draft Petition," 16.
13. Author email to Jim Milkey, February 1, 2006.

14. Jim Milkey, email, February 16, 2006.

15. Lisa Heinzerling, "Minnesota Wild," *Minnesota Law Review* 87, no. 4 (April 2003): 1139.

16. Ibid.

17. National Oceanic and Atmospheric Administration, "Trends in Atmospheric Carbon Dioxide: Data," https://www.esrl.noaa.gov /gmd/ccgg/trends/data.html.

18. Nathaniel Rich, "Losing Earth: The Decade We Almost Stopped Climate Change," *New York Times Magazine,* August 1, 2018.

19. Jeremy P. Jacobs, "Lisa Heinzerling Won't Back Down," *E&E News,* May 27, 2014, https://www.eenews.net/stories/1060000220.

20. 531 U.S. 457 (2001).

21. Richard J. Lazarus, "Advocacy Matters before the Supreme Court," *Georgetown Law Journal* 96, no. 5 (June 2008): 1493–1495.

22. Browner v. American Trucking Ass'ns, 529 U.S. 1129 (2000).

23. Sierra Club v. Morton, 401 U.S. 907 (1971).

24. Sierra Club v. Morton, 405 U.S. 727 (1972).

25. Frank Ackerman and Lisa Heinzerling, *Priceless: On Knowing the Price of Everything and the Value of Nothing* (New York: New Press, 2004).

26. Milkey, email, February 16, 2006.

27. Petition for Writ of Certiorari, Massachusetts v. EPA, 549 U.S. 497 (2007) (No. 05-1120).

28. Ibid., 3–4.

29. Milkey, email, February 16, 2006.

30. Petition for Writ of Certiorari, at 16.

31. Antonin Scalia and Bryan Garner, *Making Your Case: The Art of Persuading Judges* (St. Paul, MN: Thomson/West, 2008), 112.

32. Petition for Writ of Certiorari, at 12.

33. Ibid., 2, 4, 7, 8, 9, 12, 13, 15, 18, 20, 21, 23.

34. Jim Milkey, interview by author, July 6, 2018; Doniger, interview.

11. "Holy #@$#!"

1. Act to Establish the Department of Justice, chap. 150, 16 Stat. 162 (1870).

2. Seth P. Waxman, "Twins at Birth: Civil Rights and the Role of the Solicitor General," *Indiana Law Journal* 75, no. 4 (Fall 2000): 1300–1315.

3. H. W. Perry Jr., *Deciding to Decide: Agenda Setting in the United States Supreme Court* (Cambridge, MA: Harvard University Press, 1991), 222–245.

4. Ibid.

5. Brief for the Federal Respondent in Opposition, at 10–11, *Massachusetts*, 549 U.S. 497 (No. 05-1120).

6. Ibid., 20.

7. Reply Brief of Petitioners, at 3, *Massachusetts*, 549 U.S. 497 (No. 05-1120).

8. Ibid., 1.

9. Perry, *Deciding to Decide*, 42–43; Tony Mauro, "Court Watch: Pool Party," *National Law Journal: The Blog of the Legal Times*, September 18, 2006.

10. Adam Liptak, "Gorsuch, in Sign of Independence, Is Out of Supreme Court's Clerical Pool," *New York Times*, May 1, 2017, A22.

11. "Supreme Court Justice Stevens," C-SPAN, June 24, 2009, video, 35:23, https://www.c-span.org/video/?286081-1/supreme-court-justice-stevens.

12. Justice John Paul Stevens, interview by author, February 28, 2017.

13. Supreme Court employee, interview by author, August 1, 2018; Perry, *Deciding to Decide*, 43–44.

14. Editorial, "Warming at the Court," *Washington Post*, June 14, 2006.

15. Ibid.

16. Ibid.

17. Unnamed source, interview by author, June 27, 2018; unnamed source, interview by author, July 2, 2018.

18. Jim Milkey, email message to Bill Pardee and various *Massachusetts* petitioners, June 14, 2006.

19. Massachusetts v. EPA, 548 U.S. 903 (2006).

20. Joe Mendelson, interview by author, July 30, 2018; Joe Mendelson, email message to Ann Madding, June 20, 2017.

21. Jim Milkey, interview by author, July 6, 2018.

22. David Bookbinder, interview by author, July 17, 2018.

23. Milkey, interview, July 6, 2018.

24. Frances Beinecke correspondence to Attorney General Tom Reilly, August 25, 2005 (handwritten annotations by Jim Milkey).

12. The Lure of the Lectern

1. Supreme Court of the United States, *Rules of the Supreme Court of the United States,* Rule 28, September 12, 2017.

2. Supreme Court of the United States, "Counsel Listings, Opinions," https://www.supremecourt.gov/opinions/counsellist.aspx; Adam Feldman, "Empirical SCOTUS: Supreme Court All-Stars 2013–2017 (Corrected)," *SCOTUSBlog,* September 13, 2018, http://www.scotusblog.com/2018/09/empirical-scotus-supreme-court-all-stars-2013-2017/; Adam Feldman, "Attorneys and Firms for the 2017 Term," *Empirical SCOTUS,* May 31, 2018, https://empiricalscotus.com/2018/05/31/attorneys-firms-2017/.

3. Tony Mauro, "*Glickman v. Wileman Brothers and Elliott Inc.*: How Oral Arguments Led to a Lawsuit," in *A Good Quarrel: America's Top Legal Reporters Share Stories from Inside the Supreme Court,* ed. Timothy R. Johnson and Jerry Goldman (Ann Arbor: University of Michigan Press, 2009), 78, 92–95.

4. Lisa Heinzerling, interview by author, August 9, 2018.

5. Ibid.

6. Petition for Writ of Certiorari, at 13–24, Massachusetts v. EPA, 549 U.S. 497 (2007) (No. 05-1120).

7. Ibid., 22–26.

8. *Compare* ibid., *with* Brief for the Petitioners, at I, Massachusetts v. EPA, 549 U.S. 497 (No. 05-1120); David Doniger, interview by author, April 26, 2017.

9. *Massachusetts* Petitioners, "Draft Brief 7/28" (unpublished brief, July 28, 2006), 51–56, Microsoft Word file.

10. Joe Mendelson, memorandum to Lisa Heinzerling et al., August 1, 2006; Lisa Heinzerling, memorandum to Small Group, August 3, 2006; Jim Milkey, email message to Lisa Heinzerling, August 7, 2006.

11. Lisa Heinzerling, email message to David Bookbinder and various *Massachusetts* petitioners, August 18, 2006.

12. Lisa Heinzerling, email message to various *Massachusetts* petitioners, August 24, 2006.

13. Marc Melnick, email message to Jim Milkey and various *Massachusetts* petitioners, August 25, 2006; Jim Milkey, email message to Lisa Heinzerling and various *Massachusetts* petitioners, August 25, 2006.

14. David Doniger, email message to Lisa Heinzerling and various *Massachusetts* petitioners, August 25, 2006.

15. Heinzerling, email, August 24, 2006.

16. *Massachusetts* Petitioners, "Draft Brief 8/24" (unpublished brief, August 24, 2006), 8, 32, Microsoft Word file.

17. Brief for the Petitioners, at 48.

18. *Massachusetts* Petitioners, "Draft Brief 8/28" (unpublished brief, August 28, 2006), 50, Microsoft Word file.

19. Milkey, email, August 25, 2006.

20. Jim Milkey, email message to author, July 7, 2018.

21. David Bookbinder, interview by author, July 17, 2018.

22. David Doniger, interview by author, April 26, 2017.

23. Ibid.

24. Jim Milkey, email message to Peter Lehner, September 29, 2006.

25. Jim Milkey, interview by author, February 18, 2015; Jim Milkey, interview by author, July 6, 2018; Bookbinder, interview; Milkey, email, September 29, 2006.

26. Milkey, email, September 29, 2006.

27. David Bookbinder, email message to Jim Milkey, September 30, 2006.

28. Jim Milkey, email message to David Bookbinder, October 2, 2006.

29. Ibid.

30. Ibid.

31. Ibid.

32. Ibid.; Milkey, interview, July 6, 2015; Bookbinder, interview.

33. Lisa Heinzerling, email message to various *Massachusetts* petitioners, October 4, 2006.

34. Ibid.

35. Heinzerling, interview.

36. United States Court of Appeals District of Columbia Circuit, "Judge Cornelia T. L. Pillard," https://www.cadc.uscourts.gov/internet/home .nsf/Content/VL+-+Judges+-+NP; "Remarks by the President on the Nominations to the U.S. Court of Appeals for the District of Co-lumbia Circuit," C-SPAN, June 4, 2013, video, https://www.c-span.org /video/?313153-1/president-obama-judicial-nominations; Nevada v. Hibbs, 538 U.S. 721 (2003).

37. Heinzerling, interview.

38. Brief for the Federal Respondent, at 10–20, *Massachusetts*, 549 U.S. 497 (No. 05-1120).

39. Ibid., 14–18.
40. Ibid., 14.
41. Ibid., 33–35.
42. Ibid., 42.
43. Ibid., 20–25.
44. Ibid., 8.
45. Ibid., 42–43, 45–50.
46. Joint Brief of Industry Intervenor-Respondents, Massachusetts v. EPA, 415 F.3d 50 (D.C. Cir. 2005) (No. 03-1361).
47. Supreme Court of the United States Docket, No. 05-1120, https://www.supremecourt.gov/Search.aspx?FileName=/docketfiles\05-1120.htm.
48. Brief for the Respondent CO_2 Litigation Group, at 6–39, *Massachusetts,* 549 U.S. 497 (No. 05-1120); Brief for the Respondent States of Michigan, North Dakota, Utah, South Dakota, Alaska, Kansas, Nebraska, Texas, and Ohio, at 11–23, *Massachusetts,* 549 U.S. 497 (No. 05-1120).
49. Brief for Respondent Utility Air Regulatory Group, at 9–25, *Massachusetts,* 549 U.S. 497 (No. 05-1120).
50. Brief for Respondents Alliance of Automobile Manufacturers, Engine Manufacturers Association, National Automobile Dealers Association, Truck Manufacturers Association, at 43–50, *Massachusetts,* 549 U.S. 497 (No. 05-1120).
51. Supreme Court of the United States Docket, No. 05-1120.
52. Supreme Court of the United States, *Rules of the Supreme Court of the United States,* Rule 33, March 14, 2005.
53. Reply, *Massachusetts,* 549 U.S. 497 (No. 05-1120).
54. Ibid., 1, 22.

13. Moots

1. Jim Milkey, interview by author, July 6, 2018; David Bookbinder, interview by author, July 17, 2018.
2. Milkey, interview, July 6, 2018.
3. Ibid.
4. Supreme Court of the United States, *Guide for Counsel in Cases to Be Argued before the Supreme Court of the United States,* November 13, 2017, 6.
5. Ibid.

6. Transcript of Oral Argument, at 13, Dep't of Health and Human Servs. v. Florida, 567 U.S. 519 (2012) (No. 11-398).

7. *Guide for Counsel,* 11. For another example, see Transcript of Oral Argument, at 53, Carpenter v. United States, 138 S. Ct. 2206 (2018) (No. 16-402).

8. David Doniger, interview by author, April 26, 2017.

9. Jim Milkey, email message to author, February 20, 2015.

10. Milkey, interview, July 6, 2018; Jim Milkey, email message to author, July 17, 2019.

11. Jim Milkey, "Oral Argument Binder" (unpublished binder, November 2006), Microsoft Word file.

12. Ibid.

13. Ibid.

14. Jim Milkey, interview by author, February 18, 2015.

15. Ibid.

16. Doniger, interview.

17. Ibid.

18. Bookbinder, interview.

19. The author at the time served as the faculty co-director of the Supreme Court Institute and participated in the *Massachusetts* moot court.

20. Georgetown law student author, "Notes from Georgetown University Law Center Moot" (unpublished notes, November 17, 2006), Microsoft Word file.

21. Ibid.

22. Doniger, interview.

23. Bookbinder, interview.

24. Ibid.

25. Milkey, interview, July 6, 2018; Milkey interview, February 18, 2015; Bookbinder, interview.

26. Bookbinder, interview.

27. Milkey, interview, July 6, 2018.

28. Ibid.

14. Seventy-Four Inches

1. Jim Milkey, email message to author, March 8, 2018; Gregory Garre, email message to author, March 9, 2018; *Confirmation Hearings on Federal Appointments before the S. Comm. on the Judiciary,*

110th Cong. 623 (2008) (statement of Gregory G. Garre, Nominee to be Solicitor General of the United States).

2. Supreme Court Marshal's Office, email message to author, April 12, 2018.

3. Ibid.

4. Fred J. Maroon and Suzy Maroon, *The Supreme Court of the United States* (New York: Thomasson-Grant and Lickle, 1996), 17–33; Supreme Court Historical Society, "Homes of the Court," https://www.supremecourthistory.org/history-of-the-court/home-of -the-court.

5. Maroon and Maroon, *The Supreme Court of the United States,* 39; Supreme Court Historical Society, "Homes of the Court."

6. Supreme Court Marshal's Office, interview by author and tour of Supreme Court courtroom, April 9, 2018.

7. Ibid.

8. Ibid.

9. Jim Milkey, interview by author, July 6, 2018.

10. Jim Milkey, interview by author, February 18, 2015; author, personal recollections from argument morning at the Supreme Court, November 29, 2006.

11. Milkey, interview, July 6, 2018.

12. Former EPA employee, email message to author, June 14, 2018.

13. Milkey, email, March 8, 2018.

14. Supreme Court employee, email message to author, May 4, 2015; Maroon and Maroon, *The Supreme Court of the United States,* 134; Supreme Court Historical Society, "Homes of the Court"; Supreme Court of the United States, "Supreme Court Building— Building History," https://www.supremecourt.gov/about /buildinghistory.aspx.

15. Supreme Court employee, email message to author, March 29, 2019.

16. Maroon and Maroon, *The Supreme Court of the United States,* 134–152; Office of the Curator of the Supreme Court of the United States, "Courtroom Friezes: South and North Walls, Information Sheet," May 8, 2003, https://www.supremecourt.gov/about /northandsouthwalls.pdf; Office of the Curator of the Supreme Court of the United States, "Courtroom Friezes: East and West Walls, Information Sheet," October 1, 2010, https://www.supremecourt.gov /about/eastandwestwalls.pdf.

17. David C. Frederick, "Supreme Court Advocacy in the Early Nineteenth Century," *Journal of Supreme Court History* 30, no. 1 (March 2005): 1.

18. "Reed in Collapse; AAA Cases Halted," *New York Times,* December 11, 1935, 1, 9; John G. Roberts Jr., "Oral Advocacy and the Re-Emergence of a Supreme Court Bar," *Journal of Supreme Court History* 30, no. 1 (March 2005): 72–73.

19. Supreme Court of the United States, "The Court and Its Traditions," https://www.supremecourt.gov/about/traditions.aspx; Willard L. King, "Melville Weston Fuller: 'The Chief' and the Giants on the Court," *American Bar Association Journal* 36, no. 4 (April 1950): 349.

20. Linda Greenhouse, "David H. Souter: Justice Unbound," *New York Times,* May 2, 2009.

21. Planned Parenthood v. Casey, 505 U.S. 883 (1992).

22. Lawrence v. Texas, 539 U.S. 558 (2003).

23. Atkins v. Virginia, 536 U.S. 304 (2002); Roper v. Simmons, 543 U.S. 551 (2005).

24. Supreme Court employee, email, May 4, 2015.

25. Mark Tushnet, "Themes in Warren Court Biographies," *New York University Law Review* 70, no. 3 (June 1995): 763.

15. A Hot Bench

1. Timothy R. Johnson, email message to author, January 22, 2019; Timothy R. Johnson, Ryan C. Black, and Ryan J. Owens, "Justice Scalia and Oral Arguments at the Supreme Court," in *The Conservative Revolution of Antonin Scalia,* ed. David A. Schultz and Howard Schweber (Lanham, MD: Lexington Books, 2018), 253.

2. Paul Clement, "Supreme Court Bar Memorial for Justice Antonin Scalia," C-SPAN, November 4, 2016, video, 39:30–40:10, https://www.c-span.org/video/?417972-1/supreme-court-honors-life-justice-antonin-scalia&start=621; Transcript of Oral Argument, Hodel v. Irving, 481 U.S. 704 (1987) (No. 85-637).

3. John Calvin Jeffries, *Justice Lewis F. Powell, Jr.* (New York: Fordham University Press, 2001), 534.

4. Adam Liptak, "Vote Trading Is Not the Court's Way," Week in Review, *New York Times,* May 16, 2010, 1, 14.

5. Justice Antonin Scalia, interview by Susan Swain, C-SPAN, June 19, 2009, https://www.c-span.org/video/?286079-1/supreme-court-justice -scalia&start=955.

6. Joan Biskupic, *American Original: The Life and Constitution of Supreme Court Justice Antonin Scalia* (New York: Sarah Crichton Books, Farrar, Straus and Giroux, 2009), 122–29; Evan Thomas, *First—: Sandra Day O'Connor* (New York: Random House, 2019), 302.

7. Chief Justice John Roberts, "The Supreme Court: Home to America's Highest Court," C-SPAN, December 20, 2010, https://www.c-span.org /video/?297213-1/the-supreme-court-home-americas-highest-court -2010-edition&start=2468, at 41:26.

8. Paul Clement, "Arguing before Justice Scalia," *New York Times,* February 17, 2016 (quoting Morrison v. Olson, 487 U.S. 654, 698 (1988) (Scalia, J., dissenting)).

9. See, e.g., Transcript of Oral Argument, at 14, Missouri v. Jenkins, 515 U.S. 70 (1995) (No. 93-1823); Transcript of Oral Argument, at 19, United States v. Mezzanatto, 513 U.S. 196 (1995) (No. 93-1340).

10. Transcript of Oral Argument, at 3, Massachusetts v. EPA, 549 U.S. 497 (2007) (No. 05-1120).

11. Jim Milkey, "Oral Argument Binder" (unpublished binder, November 2006), Microsoft Word file.

12. Antonin Scalia, "The Doctrine of Standing as an Essential Element of the Separation of Powers," *Suffolk University Law Review* 17, no. 4 (1983): 897.

13. Transcript of Oral Argument, *Massachusetts,* 4.

14. Ibid.

15. Ibid., 5

16. Ibid.

17. Ibid., 6.

18. Ibid.

19. Jim Milkey, "Oral Argument Binder."

20. Transcript of Oral Argument, *Massachusetts,* 5, 13.

21. Ibid., 6–7.

22. Ibid., 10.

23. Ibid., 10–11.

24. Ibid., 11–12.

25. Ibid., 11–13.

26. Ibid., 12–13.

27. Ibid., 13.
28. Ibid., 14–15.
29. Jim Milkey, interview by author, July 6, 2018.
30. Transcript of Oral Argument, *Massachusetts,* 16.
31. Milkey, interview, July 6, 2018.
32. Transcript of Oral Argument, *Massachusetts,* 17.
33. Ibid., 18.
34. Ibid., 19.
35. Ibid.
36. Ibid., 20.
37. Ibid., 20–21.
38. Ibid., 22.
39. Ibid., 23.
40. Ibid., 23–24.
41. Ibid., 22.
42. Ibid.
43. Joe Mendelson, interview by author, July 30, 2018.
44. Transcript of Oral Argument, *Massachusetts,* 25.

16. Red Lights

1. The United States Department of Justice, "Solicitor General: Gregory G. Garre," updated October 31, 2014, https://www.justice .gov/osg/bio/gregory-g-garre; Linda Greenhouse, "Bush and First Lady Visit Rehnquist's Coffin at Court," *New York Times,* September 7, 2005, A14.

2. John G. Roberts Jr., "In Tribute to William H. Rehnquist," *Columbia Law Review* 106, no. 3 (April 2006): 487; Kerri Martin Bartlett, "Memories of a Modest Man: A Tribute to Chief Justice William H. Rehnquist," *Columbia Law Review* 106, no. 3 (April 2006): 490; Michael K. Young, "Croquet, Competition, and the Rules: A More Personal Reflection on the Jurisprudence of Chief Justice William H. Rehnquist," *Columbia Law Review* 106, no. 3 (April 2006): 498; Gregory G. Garre, "Commencement Address," George Washington University Law School, May 15, 2016.

3. Gregory Garre, interview by author, June 3, 2015.

4. Margaret Meriwether Cordray and Richard Cordray, "The Solicitor General's Changing Role in Supreme Court Litigation," *Boston*

College Law Review 51, no. 5 (November 2010): 1354; Lincoln Caplan, *The Tenth Justice: The Solicitor General and the Rule of Law* (New York: Alfred A. Knopf, 1987), 1.

5. Supreme Court of the United States, "About the Court, Building Features," https://www.supremecourt.gov/about/buildingfeatures .aspx.

6. Debra Cassens Weiss, "SG Dumped Traditional Morning Coat, Wore Pantsuit of Unknown Design," *ABA Journal,* September 11, 2009; Marcia Coyle, "Morning Coats and First Arguments: Female SCOTUS Lawyers on Breaking Barriers," *National Law Journal,* October 6, 2017.

7. Gregory Garre, email message to author, March 9, 2018.

8. Gregory Garre, email message to author, July 18, 2019.

9. Garre, interview.

10. Ibid.

11. Transcript of Oral Argument, at 29–31, 40, Massachusetts v. EPA, 549 U.S. 497 (2007) (No. 05-1120); RonNell Andersen Jones and Aaron L. Nieldon, "Clarence Thomas the Questioner," *Northwestern University Law Review Online* 111 (2017): 197.

12. Transcript of Oral Argument, at 25.

13. Ibid.

14. Ibid., 27.

15. Ibid., 28.

16. Ibid., 28–29; Oyez, Massachusetts v. EPA, Oral Argument (November 29, 2006), https://www.oyez.org/cases/2006/05-1120.

17. Transcript of Oral Argument, at 28–29, Massachusetts v. EPA.

18. Ibid., 41, 43, 50–51.

19. Ibid., 52.

20. Ibid.

21. Ibid., 51.

22. Ibid., 53.

23. Jim Milkey, interview by author, July 6, 2018.

24. Transcript of Oral Argument, at 53, Massachusetts v. EPA.

25. Ibid., 56.

26. Milkey, interview, July 6, 2018.

27. Cathie Jo Martin, interview by author, February 11, 2019.

28. Jim Milkey, interview by author, February 18, 2015; David Doniger, email message to author, January 20, 2019.

29. Milkey, interview, July 6, 2018; Jim Milkey, email message to author, March 8, 2018.

17. The Conference

1. Dahlia Lithwick, "Bonus Round: What to Make of These Astronomical Supreme Court Signing Bonuses?," *Slate,* March 10, 2007, https://slate.com/news-and-politics/2007/03/what-to-make-of-those-astronomical-supreme-court-signing-bonuses.html.
2. Tony Mauro, "$400K for SCOTUS Clerks: A Bonus Too Far?," *National Law Journal,* November 14, 2018.
3. Federal Judicial Center, "Judicial Salaries: Supreme Court Justices," https://www.fjc.gov/history/judges/judicial-salaries-supreme-court-justices.
4. Ted Cruz (former law clerk to Chief Justice William Rehnquist), email to unnamed former Supreme Court law clerks, April 16, 1998.
5. William H. Rehnquist, *The Supreme Court* (New York: Alfred A. Knopf, 2001), 240; former Supreme Court employee, email message to author, January 23, 2019; former Supreme Court employee, email message to author, January 25, 2019.
6. Wikipedia, "List of Law Clerks of the Supreme Court of the United States," https://en.wikipedia.org/wiki/List_of_law_clerks_of_the_Supreme_Court_of_the_United_States.
7. Adam Liptak and Emmarie Huetteman, "Justice Antonin Scalia Honored at Supreme Court," *New York Times,* February 20, 2016, A12; Lauren Markoe, "Scalia Mourned by Thousands at Supreme Court," *Washington Post,* February 19, 2016; Robert Barnes and Cortlynn Stark, "John Paul Stevens Remembered for 'Deep Devotion' to Law and Justice," *Washington Post,* July 23, 2019, A3.
8. Stanley Kay, "The Highest Court in the Land," *Sports Illustrated,* July 30, 2018, 66.
9. Rehnquist, *The Supreme Court,* 253.
10. Fred J. Maroon and Suzy Maroon, *The Supreme Court of the United States* (New York: Thomasson-Grant and Lickle, 1996), 77, 89; Supreme Court Historical Society, "Homes of the Court," https://supremecourthistory.org/history-of-the-court/home-of-the-court/.
11. Wikipedia, "List of Law Clerks."

12. Chief Justice John G. Roberts, "In Tribute: Justice Anthony M. Kennedy," *Harvard Law Review* 132, no. 1 (November 2018): 1, 24–27.

13. Adam Liptak, "For a Collegial Court, Justices Lunch Together, and Forbid Talk of Cases," *New York Times,* June 1, 2016; Maria Godoy, "For a Cordial Supreme Court, Keep the Food and Wine Coming," NPR, June 3, 2016, https://www.npr.org/sections/thesalt/2016/06/03 /480503335/for-a-cordial-supreme-court-keep-the-food-and-wine -coming; Terry Stephan, "A Justice for All," *Northwestern Magazine,* Spring 2009.

14. Jean Edward Smith, *John Marshall: Definer of a Nation* (New York: Henry Holt, 1996), 378.

15. Ronald D. Rotunda, *John Marshall and the Cases That United the States of America* (Northport, NY: Twelve Tables Press, 2018), 297.

16. Godoy, "For a Cordial Supreme Court."

17. Liptak, "For a Collegial Court"; Deborah Simmons, "Antonin Scalia: Foodie, 'Man of Many Appetites,'" *Washington Times,* February 19, 2016; former Supreme Court employee, email message to author, March 25, 2018.

18. Justice Clarence Thomas, "The Supreme Court: Home to America's Highest Court," C-SPAN, December 20, 2010, https://www.c-span.org /video/?297213-1/the-supreme-court-home-americas-highest-court -2010-edition&start=2468, at 52:59.

19. Supreme Court employee, email message to author, February 14, 2019; Maroon and Maroon, *The Supreme Court,* 114–115; Supreme Court Public Information Office, email message to author, August 28, 2018; Supreme Court Public Information Office, interview by author, July 24, 2018; "The Supreme Court: Home to America's Highest Court," at 1:01:37.

20. Public Information Office, interview.

21. Clare Cushman, "Rookie on the Bench: The Role of the Junior Justice," *Journal of Supreme Court History* 32, no. 3 (November 2007): 289–290.

22. Dred Scott v. Sandford, 60 U.S. (19 How.) 393, 529 (1857) (McLean, J., dissenting).

23. Maroon and Maroon, *The Supreme Court,* 114–115, 158–159; Supreme Court Public Information Office, email message to author, September 10, 2018.

24. Georgia v. Tennessee Cooper Co., 200 U.S. 230 (1907).

25. Supreme Court employee, interview by author, August 1, 2018; "The Supreme Court: Home to America's Highest Court," at 1:01:37.

26. "October Term 2006," *Journal of the Supreme Court,* June, 2007, 392–395, https://www.supremecourt.gov/orders/journal/jnl06.pdf.

27. Rehnquist, *The Supreme Court,* 254–255.

28. John G. Roberts Jr., "Article III Limits on Statutory Standing," *Duke Law Journal* 42, no. 6 (April 1993): 1219–1232; *Confirmation Hearing on the Nomination of John G. Roberts, Jr. to Be Chief Justice of the United States,* 109th Cong. 155–156, 342 (2005) (testimony of John G. Roberts Jr.).

29. Massachusetts v. EPA, 549 U.S. 497, 535 (2007) (internal quotations omitted).

30. Kelly Lynn Brown and Rebecca Agule, "Scalia Speaks in Ames, Scolds Aggressive Student," *Harvard Law Record,* December 7, 2006, https://archive.li/k1DWn.

31. Supreme Court employee, interview, August 1, 2018.

32. Supreme Court employee, interview by author, August 13, 2018.

33. Stephen G. Breyer, *Breaking the Vicious Circle* (Cambridge, MA: Harvard University Press, 1992).

34. Whitman v. American Trucking Assns., Inc., 531 U.S. 457, 490 (2001) (Breyer, J., concurring in part and concurring in the judgment).

35. Tara Leigh Grove, "Book Review: The Supreme Court's Legitimacy Dilemma," *Harvard Law Review* 132, no. 9 (2019): 2243; National Federation of Independent Business v. Sebelius, 567 U.S. 519 (2012).

36. 505 U.S. 833 (1992); Evans Rowl and Robert Novak, "Justice Kennedy's Flip," *Washington Post,* September 4, 1992.

37. Justice John Paul Stevens, interview by author, February 28, 2017.

38. Abe Fortas, "Chief Justice Warren: The Enigma of Leadership," *Yale Law Journal* 84, no. 3 (1975): 405.

39. Jeffrey Rosen, "The Agonizer," *New Yorker,* November 11, 1996, 85.

40. Stevens, interview by author.

41. Civil Rights Act of 1991, Pub. L. No. 102-166, 105 Stat. 1071 (1991).

42. Alaska Dep't of Envtl. Conserv. v. EPA, 540 U.S. 461, 502 (2004) (Kennedy, J., dissenting); Solid Waste Agency of N. Cook County v. U.S. Army Corps of Eng'rs, 531 U.S. 159 (2001).

43. Evan Thomas, *First: Sandra Day O'Connor* (New York: Random House, 2019), 446.

44. David A. Kaplan, *The Most Dangerous Branch* (New York: Crown Publishing, 2018), 158.
45. Massachusetts v. EPA, 549 U.S. 497 (2007); Weyerhaeuser Co. v. Ross-Simmons Hardware Lumber Co., 549 U.S. 312 (2007); KSR International Co. v. Teleflex Inc., 550 U.S. 398 (2007).
46. Gonzales v. Duenas Alvarez, 549 U.S. 183, 192 (2007); Rockwell International Corp v. United States, 549 U.S. 457, 479 (2007); Watters v. Wachovia Bank, 550 U.S. 1, 22 (2007); Bell Atlantic v. Twombly, 550 U.S. 544, 570 (2007); Ledbetter v. Goodyear Tire & Rubber Co., Inc., 550 U.S. 618, 643 (2007); Parents Involved in Community Schools v. Seattle School District, 551 U.S. 701, 798 (2007).
47. *Weyerhaeuser,* 549 U.S., at 312.
48. KSR International Co., 550 U.S., at 398.
49. Richard J. Lazarus, "Back to 'Business' at the Supreme Court: The 'Administrative Side' of Chief Justice Roberts," *Harvard Law Review Forum* 129, no. 1 (November 2015): 63.

18. A Bow-Tied Jedi Master

1. Jeffrey Rosen, "The Dissenter, Justice John Paul Stevens," *New York Times Magazine,* September 23, 2007; Akin Gump Strauss Hauer & Feld, "End of Term Statistics and Analysis—October Term 2005," June 29, 2006, https://www.scotusblog.com/archives /EndofTermAnalysis.pdf.
2. Akin Gump Strauss Hauer & Feld, "End of Term Statistics and Analysis—October Term 2006," June 28, 2007, https://www .scotusblog.com/archives/SuperStatPack.pdf.
3. Gonzales v. Carhart, 550 U.S. 124 (2007).
4. Parents Involved in Community Schools v. Seattle School District No. 1, 551 U.S. 701 (2007).
5. Oyez, "Parents Involved in Community Schools v. Seattle School District No. 1, Opinion Announcement, June 28, 2007," https://www .oyez.org/cases/2006/05-908.
6. Fred J. Maroon and Suzy Maroon, *The Supreme Court of the United States* (New York: Thomasson-Grant and Lickle, 1996), 112–113.
7. "Supreme Court Justice Stevens," C-SPAN, June 24, 2009, video, https://www.c-span.org/video/?286081-1/supreme-court-justice

-stevens; Terry Stephan, "A Justice for All," *Northwestern Magazine,* Spring 2009, 4.

8. Justice John Paul Stevens, *The Making of a Justice: My First 94 Years* (New York: Little Brown, 2019), 10–11; Bill Barnhart and Gene Schlickman, *John Paul Stevens: An Independent Life* (DeKalb: Northern Illinois University Press, 2010), 26–27; Stephan, "A Justice for All," 1.

9. Stevens, *The Making of a Justice,* 18; Ed Sherman, "Did Babe Ruth's Called Shot Happen?," *Chicago Tribune,* March 28, 2014; Rosen, "The Dissenter," 50.

10. Barnhart and Schlickman, *John Paul Stevens,* 32–35; Rosen, "The Dissenter"; Stevens, *The Making of a Justice,* 24–25.

11. Stevens, *The Making of a Justice,* 527; "Supreme Court Justice Stevens," C-SPAN, 12:38; Barnhart and Schlickman, *John Paul Stevens,* 135.

12. John M. Ferren, *Salt of the Earth, Conscience of the Court: The Story of Justice Wiley Rutledge* (Chapel Hill: University of North Carolina Press, 2004).

13. "Supreme Court Justice Stevens," C-SPAN, 12:38; former Supreme Court employee, email message to author, March 25, 2018.

14. 335 U.S. 188 (1948).

15. 542 U.S. 466, 477 & n.7 (2004).

16. 548 U.S. 557, 618 & n.46 (2005), citing *Yamashita,* 327 U.S. 1, 44 (1946) (Murphy & Rutledge, JJ., dissenting).

17. Former Supreme Court employee, email message to author, March 25, 2018; Barnhart and Schlickman, *John Paul Stevens,* 3–4, 200.

18. Wikipedia, "List of Law Clerks of the Supreme Court of the United States (Seat 4)," https://en.wikipedia.org/wiki/List_of_law_clerks_of _the_Supreme_Court_of_the_United_States_(Seat_4); Jeff Pearlman, "Jamal Greene," July 31, 2017, http://www.jeffpearlman.com/jamal -greene/; Munger Tolles & Olson, "Chad Golder," https://www.mto .com/lawyers/chad-golder; University of Michigan Law School, "Bagley, Nicholas," https://www.law.umich.edu/FacultyBio/Pages /FacultyBio.aspx?FacID=nbagley; Georgia State University College of Law, "Lauren Sudeall," https://law.gsu.edu/profile/lauren-sudeall/.

19. Barnhart and Schlickman, *John Paul Stevens,* 3–4.

20. Former Supreme Court employee, email, March 25, 2018.

21. Pamela Harris, "The Importance of Stevens' Good Manners," *SCO-TUSblog,* April 26, 2010, https://www.scotusblog.com/2010/04/the -importance-of-stevens-good-manners/.

22. Barnhart and Schlickman, *John Paul Stevens*, 200–201, 205–206, 229; Rosen, "The Dissenter."

23. Rosen, "The Dissenter."

24. Ibid.; "Supreme Court Justice Stevens," C-SPAN, 12:20, 18:50.

25. "Supreme Court Justice Stevens," C-SPAN, 6:38.

26. "Conversation with Justice John Paul Stevens," C-SPAN, July 19, 2007, video, 7:35, 7:43, https://www.c-span.org/video/?200035-2/conversation-justice-john-paul-stevens&start=452.

27. Justice John Paul Stevens, interview by author, February 28, 2017.

28. Ibid.

29. Ibid.

30. Ibid.

31. 547 U.S. 715 (2006); Richard J. Lazarus, "Back to 'Business' at the Supreme Court: The "Administrative Side' of Chief Justice Roberts," *Harvard Law Review Forum* 129, no. 2 (2015): 33, 66.

32. David A. Kaplan, *The Most Dangerous Branch: Inside the Supreme Court's Assault on the Constitution* (New York: Crown, 2018), 158.

33. Stevens, *The Making of a Justice*, 464.

34. John F. Manning, "In Memoriam: Justice Antonin Scalia," *Harvard Law Review* 130, no. 1 (2016): 14–15; Kannon K. Sanmugam, "Justice Scalia: A Personal Remembrance," *Journal of Supreme Court History* 41, no. 3 (2016): 252–253.

35. "Conversation with Justice John Paul Stevens," C-SPAN, 5:07; William H. Rehnquist, *The Supreme Court* (New York: Alfred A. Knopf, 2001), 264.

36. Rehnquist, *The Supreme Court*, 264; see, for example, Justice Anthony M. Kennedy to Justice O'Connor, memorandum, June 11, 1992, and Justice Anthony M. Kennedy to Justice Scalia, memorandum, May 28, 1992, both box 1405, Harry A. Blackmun Papers, Manuscript Division, Library of Congress.

37. See, for example, Justice David H. Souter to the Chief Justice, memorandum, May 17, 1994, and Justice Anthony M. Kennedy to the Chief Justice, memorandum, May 16, 1994, boxes 1408 and 1405, Harry A. Blackmun Papers, Manuscript Division, Library of Congress.

38. Supreme Court employee, interview by author, August 13, 2018.

39. Ibid.

40. Ibid.

41. 206 U.S. 230 (1907); Transcript of Oral Argument, at 14–15, Massachusetts v. EPA, 549 U.S. 497 (2007) (No. 05-1120).

42. Jim Milkey, email message to various *Massachusetts* petitioners, December 10, 2003; Massachusetts v. EPA, 549 U.S. 497, 539 (2007) (Roberts, C. J., dissenting).

43. *Massachusetts,* 549 U.S. 497, 516–520; Rosen, "The Dissenter."

44. *Massachusetts,* 549 U.S. 497, 534–535.

45. Supreme Court employee, interview, August 13, 2018.

46. *Massachusetts,* 549 U.S. 497, 534–535.

47. Ibid., 520.

48. Ibid., 504–505.

49. Justice John Paul Stevens, interview by author, February 28, 2017.

50. *Massachusetts,* 549 U.S. 497, 528, 529, 532.

19. Two Boxes

1. Supreme Court employee, interview by author, August 1, 2018.

2. Ibid.

3. Robert Barnes, interview by author, February 9, 2019.

4. "October Term 2006," *Journal of the Supreme Court,* June 2007, 797, https://www.supremecourt.gov/orders/journal/jnl06.pdf.

5. Jim Milkey, email message to author, November 19, 2018; Jim Milkey, email message to author, November 20, 2018.

6. Lisa Heinzerling, interview by author, August 9, 2018.

7. David Doniger, interview by author, April 26, 2017.

8. David Bookbinder, interview by author, July 17, 2018.

9. Joe Mendelson, interview by author, July 30, 2018.

10. Gregg Garre, email to author, July 22, 2019.

11. "October Term 2006," 810; Oyez, "Environmental Defense v. Duke Energy Corporation: Opinion Announcement, April 02, 2007," https://www.oyez.org/cases/2006/05-848.

12. "October Term 2006," 810; Oyez, "Massachusetts v. Environmental Protection Agency: Opinion Announcement, April 02, 2007," https://apps.oyez.org/player/#/roberts2/opinion_announcement _audio/21938.

13. Oyez, "Massachusetts v. Environmental Protection Agency: Opinion Announcement, April 02, 2007," https://apps.oyez.org/player/# /roberts2/opinion_announcement_audio/21938.

14. Gregory G. Garre, "Commencement Address," George Washington University School of Law (May 15, 2016).

15. Supreme Court Public Information Office, email to author, March 11, 2019; "The Supreme Court: Home to America's Highest Court," C-SPAN, December 20, 2010, https://www.c-span.org/video/ ?297213-1/the-supreme-court-home-americas-highest-court-2010 -edition&start=2468, at 1:16:40.

16. Milkey, email, November 19, 2018; Milkey, email, November 20, 2018; Jim Milkey, "U.S. Supreme Court Rules in Favor of Massachusetts in Global Warming Case" (unpublished press release, March 19, 2007), Microsoft Word file; Jim Milkey, "U.S. Supreme Court Rules EPA Has Authority under Clean Air Act to Regulate Greenhouse Gases" (unpublished press release, March 19, 2007), Microsoft Word file; Jim Milkey, "U.S. Supreme Court Issues Ruling in Global Warming Case" (unpublished press release, March 19, 2007), Microsoft Word file.

17. Bookbinder, interview; Tom Pelton, "Justices Rebuke Bush on Climate," *Baltimore Sun,* April 3, 2007.

18. Doniger, interview.

19. Lisa Heinzerling, "High Court Rebukes Bush on Energy, Environment," interview by Elizabeth Shogren, *All Things Considered,* NPR, April 2, 2007, audio, 00:57, https://www.npr.org/templates/story /story.php?storyId=9293462.

20. Heinzerling, interview.

21. Mendelson, interview, July 30, 2018.

22. International Center for Technology Assessment, "Supreme Court Finds that Bush Administration EPA Illegally Resisted Efforts to Regulate Global Warming Pollution," press release, April 2, 2007.

23. Joe Mendelson, email message to author, June 20, 2017.

24. Linda Greenhouse, "Justices Say E.P.A. Has Power to Act on Harmful Gases: Agency Can't Avoid Its Authority—Rebuke to Administration," *New York Times,* April 3, 2007, A1.

25. Jess Bravin, "Court Rulings Could Hit Utilities, Auto Makers; White House Strategy toward CO_2 Emissions Is Faulted by Justices," *Wall Street Journal,* April 3, 2007, A1.

26. Editorial, "Jolly Green Justices," *Wall Street Journal,* April 3, 2007, A14.

27. Earthjustice, advertisement, *New York Times,* April 5, 2007, A13.

28. Jeffrey Holmstead, interview by author, August 13, 2018.

29. Howard Kohn, "A Law against Greenhouse Gases: How Two Local Attorneys Made History," *TakomaVoice*, May 10, 2010.

30. 347 U.S. 483 (1954).

31. 5 U.S. (1 Cranch) 137 (1803).

32. "Text of Supreme Court Decision Outlawing Negro Segregation in the Public Schools," *New York Times*, May 18, 1954, 15; "Opinion on D.C. Schools," *Washington Post and Times Herald*, May 18, 1954, A1; "Supreme Court's Decision in U.S. School Segregation Cases," *Chicago Daily Tribune*, May 18, 1954, 8.

33. Brown v. Board, 346 U.S. 483 (1955).

34. Massachusetts v. EPA, 549 U.S. 497, 534 (2007).

35. President George W. Bush, "President Bush Discusses CAFE and Alternative Fuel Standards," May 14, 2007, https://georgewbush -whitehouse.archives.gov/news/releases/2007/05/20070514-4.html.

36. President George W. Bush, "Briefing by Conference Call on the President's Announcement on CAFE and Alternative Fuel Standards," May 14, 2007, https://georgewbush-whitehouse.archives.gov/news /releases/2007/05/20070514-6.html.

37. Amanda Little, "Bush EPA Nominee Stephen Johnson Garners Praise and Sympathy," *Grist*, March 9, 2005, https://grist.org/article/little -johnson/; President George W. Bush, "President Nominates Steve Johnson as EPA Administrator," March 4, 2005, https://georgewbush -whitehouse.archives.gov/news/releases/2005/03/20050304-2.html.

38. *EPA Approval of New Power Plants: Failure to Address Global Warming Pollutants: Hearing before the H. Comm. on Oversight and Government Reform*, 110th Cong. 25 (2007) (Statement of Stephen L. Johnson, Administrator, Environmental Protection Agency); letter from Henry A. Waxman, Chairman, H. Comm. on Oversight and Government Reform, to Stephen Johnson, Administrator, Environmental Protection Agency (March 12, 2008).

39. Letter from Waxman to Johnson.

40. Felicity Barringer, "White House Refused to Open Pollutants E-Mail," *New York Times*, June 25, 2008, A15; "Be Patient, This Gets Amazing—EPA Email," *The Daily Show with Jon Stewart*, aired June 25, 2008, on Comedy Central, 1:30, http://www.cc.com/video -clips/gp4pch/the-daily-show-with-jon-stewart-be-patient-this-gets -amazing-epa-e-mail.

41. "Regulating Greenhouse Gas Emissions under the Clean Air Act," 73 Fed. Reg. 44,354, 44,396, 44,355 (July 30, 2008).
42. James Hansen et al., "Target Atmospheric CO_2: Where Should Humanity Aim?," *Open Atmospheric Science Journal* 2 (2008): 217–231.
43. National Oceanic and Atmospheric Administration, "Trends in Atmospheric Carbon Dioxide: Data," https://www.esrl.noaa.gov /gmd/ccgg/trends/data.html.

20. Making History

1. Charlie Savage, "Shepherd of a Government in Exile," *New York Times,* November 7, 2008, A22; NRDC Action Fund, "Return of Organization Exempt from Income Tax," Internal Revenue Service Form 990 (2007) (for tax year beginning July 1, 2007 and ending June 30, 2008); "Save the Economy, and the Planet," *New York Times,* November 27, 2008, A38; Georgetown Law, "John Podesta (L '76) Delivers 2017 Graduating Class Lecture," April 18, 2017, https://www .law.georgetown.edu/news/john-podesta-l76-delivers-2017-graduating -class-lecture/.
2. John M. Broder and Andrew C. Revkin, "Hard Task for New Team on Energy and Climate," *New York Times,* December 16, 2008, A24.
3. Ibid.; "Mr. Obama's Green Team," *New York Times,* December 13, 2008, A20.
4. CQ Transcript Wire, "Steven Chu Confirmation Hearing," January 14, 2009, http://www.washingtonpost.com/wp-dyn/content/article/2009/01 /19/AR2009011901107.html.
5. Andrew C. Revkin and Cornelia Dean, "For Science Adviser, Dogged Work against Global Perils," *New York Times,* December 22, 2008, D3.
6. "Save the Economy," *New York Times.*
7. Al Kamen, "Moving In," *Washington Post,* January 30, 2009, A17.
8. Steven Mufson, "Will Obama's Revolution Deliver Energy Independence?," *Washington Post,* April 5, 2009, B2.
9. Lorraine C. Miller, "Statistics of the Presidential and Congressional Election of November 4, 2008" (Clerk of the House of Representatives, 2009), 75; Carl Hulse, "What's So Super about a Supermajority?," *New York Times,* July 2, 2009, A13; "Election Results 2008: Senate—Big Board," *New York Times,* December 9, 2008.

10. John M. Broder, "Democrats Oust Longtime Leader of House Panel," *New York Times,* November 20, 2008, A1; Final Amicus Curiae Brief of John Dingell (D-Michigan) in Support of Denial of Petitioners for Review, Massachusetts v. EPA, 415 F.3d 50 (D.C. Cir. 2005) (No. 03-1361); Massachusetts v. EPA, Order Denying the Motion of John Dingell to Present Oral Argument (D.C. Cir., March 30, 2005) (No. 03-1361).

11. President Barack Obama, "First Inaugural Address," January 21, 2009, https://obamawhitehouse.archives.gov/blog/2009/01/21/president-barack-obamas-inaugural-address.

12. President Barack Obama, "Remarks on Achieving Energy Independence," January 26, 2009, transcript, https://insideclimatenews.org/sites/default/files/OBAMA.pdf.

13. United States Energy Information Administration, "International Energy Statistics," https://www.eia.gov/beta/international/data/browser/#/?c=4100000002000060000000000000g00020000000000000000001&vs=INTL.44-1-AFRC-QBTU.A&vo=0&v=H&end=2016&showdm=y; International Energy Agency, "Statistics—Atlas of Energy," http://energyatlas.iea.org/#!/tellmap/1378539487; Marcia Rocha et al., "Historical Responsibility for Climate Change—From Countries Emissions to Contributions Temperature Increase" (Climate Analytics, November 2015), 8, https://climateanalytics.org/media/historical_responsibility_report_nov_2015.pdf.

14. John M. Broder, "House Backs Bill, 219–212, to Curb Global Warming," *New York Times,* June 27, 2009, A1.

15. Joe Mendelson, email to author, July 26, 2019.

16. Jonathan Weisman and Ashley Parker, "Riding Wave of Discontent, G.O.P. Takes Senate," *New York Times,* November 5, 2014, A1; Jeff Zeleny, "G.O.P. Captures House, but Not Senate," *New York Times,* November 3, 2010, A1.

17. Ryan Lizza, "As the World Burns," *New Yorker,* October 11, 2010; Tim Dickinson, "Climate Bill, R.I.P.," *Rolling Stone,* July 21, 2010.

18. "Proposed Endangerment and Cause or Contribute Findings for Greenhouse Gases under Section 202(a) of the Clean Air Act," 74 Fed. Reg. 18,886 (April 24, 2009); John M. Broder, "E.P.A. Clears Way for Greenhouse Gas Rules," *New York Times,* April 18, 2009, A15.

19. "Light-Duty Vehicle Greenhouse Gas Emission Standards and Corporate Average Fuel Economy Standards, Final Rule," 75 Fed. Reg. 25,324, 25,326 (May 7, 2010).

20. Former EPA political appointee, interview with author, September 12, 2018.

21. "Endangerment and Cause or Contribute Findings for Greenhouse Gases under Section 202(a) of the Clean Air Act, Final Rule," 74 Fed. Reg. 66,496 (December 15, 2009).

22. "2017 and Later Model Year Light-Duty Vehicle Greenhouse Gas Emissions and Corporate Average Fuel Economy Standards," 77 Fed. Reg. 62,624 (Oct. 15, 2012); "Greenhouse Gas Emissions and Fuel Efficiency Standards for Medium and Heavy-Duty Engines and Vehicles, Phase 2," 81 Fed. Reg. 73,478 (Oct. 25, 2016).

23. Office of the Press Secretary, "President Obama Announces Historic 54.5 mpg Fuel Efficiency Standard," July 29, 2011, https://obamawhitehouse.archives.gov/the-press-office/2011/07/29/president-obama-announces-historic-545-mpg-fuel-efficiency-standard.

24. Office of the Governor, State of Massachusetts, "Governor Patrick Nominates Esteemed Pair to the Appeals Court," press release, March 6, 2009; Mass.gov, Associate Justice James R. Milkey, https://www.mass.gov/service-details/associate-justice-james-r-milkey; Jim Milkey, email message to author, October 23, 2018.

25. President Barack Obama, "Remarks by the President on Climate Change," June 25, 2013, https://obamawhitehouse.archives.gov/the-press-office/2013/06/25/remarks-president-climate-change.

26. President Barack Obama, "Remarks by the President on the American Automotive Industry," March 30, 2009, https://obamawhitehouse.archives.gov/the-press-office/remarks-president-american-automotive-industry-33009.

27. President Barack Obama, "Remarks by the President on National Fuel Efficiency Standards," May 19, 2009, https://obamawhitehouse.archives.gov/the-press-office/remarks-president-national-fuel-efficiency-standards; Jody Freeman, "The Obama Administration's National Auto Policy: Lessons from the 'Car Deal,'" *Harvard Environmental Law Review* 35, no. 2 (2011): 343–373.

28. Coral Davenport, "McConnell Wants States' Help against an Obama 'War on Coal,'" *New York Times,* March 20, 2015, A1; Ken Ward Jr., "Here's Hoping: How Coal Boosters Hold W.Va. Back," *Coal Tattoo* (blog), *Charleston Gazette-Mail: Coal Tattoo,* April 28, 2014, http://blogs.wvgazettemail.com/coaltattoo/2014/04/28/heres-hoping-how-coal-boosters-hold-w-va-back/.

29. Theda Skocpol, "Naming the Problem: What It Will Take to Counter Extremism and Engage Americans in the Fight against Global Warming" (unpublished symposium report, Harvard University, January 2013), 82–93.

30. Bruce Alpert, "Vitter Says Emails Show EPA Is 'Cozy' with Group— Sides Says Contact Was Appropriate" *New Orleans Times Picayune,* October 15, 2014.

31. John M. Broder, "5 Nations Forge Pact on Climate; Goals Go Unmet," *New York Times,* December 19, 2009, A1.

32. "Carbon Pollution Emission Guidelines for Existing Stationary Sources: Electric Utility Generating Units, Proposed Rule," 79 Fed. Reg. 34,830 (June 18, 2014); "Carbon Pollution Emission Guidelines for Existing Stationary Sources: Electric Utility Generating Units, Final Rule," 80 Fed. Reg. 64,662, 64,663 (October 23, 2015).

33. Environmental Protection Agency, "Regulatory Impact Analysis for the Clean Power Plan," Final Rule (October 23, 2015), ES-8, ES 22 T, ES-9, 3–40.

34. National Oceanic and Atmospheric Administration, "Trends in Atmospheric Carbon Dioxide: Data," https://www.esrl.noaa.gov/gmd /ccgg/trends/data.html.

35. Scott C. Doney, William M. Balch, Victoria J. Fabry, and Richard A. Feely, "Ocean Acidification: A Critical Emerging Problem for the Ocean Sciences," *Oceanography* 22, no. 4 (December 2009): 16–25.

36. Brookings Institution, "Unpacking the Paris Climate Conference: A Conversation with Todd Stern," interview by Bruce Jones, December 18, 2015, https://www.bing.com/videos/search?q=todd+stern+ paris+climate+accord&&view=detail&mid=46A2B92DE8BD4C07B8 D846A2B92DE8BD4C07B8D8&&FORM=VRDGAR.

37. "No 'Plan B' for Climate Action as There Is No 'Planet B,' Says UN Chief," *UN News,* September 21, 2014.

38. Coral Davenport, "Nations Approve Landmark Climate Deal," *New York Times,* December 13, 2015, A1.

39. U.S. Global Change Research Program, "Fourth National Climate Assessment," November 2018, 6, 12–18, 33–38, 40–42, 47–48, 64, 82–83, 94.

40. Joe Mendelson, email message to author, July 27, 2019.

41. David Bookbinder, email message to author, July 30, 2019.

42. David Doniger, email message to author, September 12, 2018.

43. David Doniger, email message to author, September 11, 2018.

Epilogue

1. Joe Mendelson, email message to author, September 15, 2018.
2. Peter Baker, "No One Will Say if Trump Denies Climate Science," *New York Times,* June 3, 2017, A1; Donald J. Trump, Twitter post, November 6, 2012, 11:15 a.m., https://twitter.com /realDonaldTrump.
3. Lisa Friedman, "Andrew Wheeler, Who Continued Environmental Rollbacks, Is Confirmed to Lead EPA," *New York Times,* March 20, 2019, A19.
4. Coral Davenport and Eric Lipton, "Choice for E.P.A. Has Led Battles to Constrain It," *New York Times,* December 8, 2016, A1; Coral Davenport, "Perry Is Chosen as Energy Chief," *New York Times,* December 14, 2016, A1.
5. "Exec. Order No. 13,789," 82 Fed. Reg. 16093 (March 31. 2017).
6. Mike Ives, "Promised Billions for Climate Change, Poor Countries Are Still Waiting," *New York Times,* September 10, 2018, A10; Brad Plumer, "U.S. Won't Actually Be Leaving the Paris Climate Deal Anytime Soon," *New York Times,* June 7, 2017, A20; U.S. Department of State, "Communication Regarding Intent to Withdraw From Paris Agreement," media note, August 4, 2017; Brad Plumer, "What to Expect as U.S. Leaves Paris Climate Accord," *New York Times,* June 1, 2017.
7. Coral Davenport, "Trump to Scrap California's Role on Car Emissions," *New York Times,* September 18, 2019, A1; California Air Resources Board, "California Moves to Ensure Vehicles Meet Existing State Greenhouse Gas Emissions Standards," press release no. 18-42, August 7, 2018, https://ww2.arb.ca.gov/news/california-moves-ensure -vehicles-meet-existing-state-greenhouse-gas-emissions-standards-0; "The Safer Affordable Fuel-Efficient (SAFE) Vehicles Rule for Model Years 2021–2026 Passenger Cars and Light Trucks," 83 Fed. Reg. 42986, 42989–42999 (Aug. 24, 2018); Brad Plumer, "How Much Car Pollution? A Whole Country's Worth," *New York Times,* August 4, 2018, A1.
8. EPA, "Repeal of the Clean Power Plan; Emissions Guidelines for Greenhouse Gas Emissions from Existing Electrical Utility Generating Units; Revisions to Emission Guidelines Implementing Regulations," June 19, 2019; Lisa Friedman, "EPA Finalizes Its Plan to Replace Obama-Era Climate Rules," *New York Times,* June 19, 2019; Lisa

Friedman, "Cost of E.P.A.'s Pollution Rules: Up to 1,400 More Deaths a Year," *New York Times,* August 22, 2018, A1.

9. Chris Mooney, "The Energy 202: Trump's Budget Seeks Cuts to Climate Research and Renewable Research Energy Programs," *Washington Post,* March 12, 2019.

10. A. Jay et al., eds., "Overview," in *Impacts, Risks, and Adaptation in the United States: Fourth National Climate Assessment,* vol. 2 (Washington, DC: U.S. Global Change Research Program, 2017), 33–36, 39, 43.

11. Ibid., 26, 34, 50.

12. Ibid., 33–58, 72–102.

13. Clean Air Council v. Pruitt, 862 F.3d 1 (D.C. Cir. 2017) (methane emission regulations); Natural Resources Defense Council v. National Highway Traffic Administration, 894 F.3d 92 (2d Cir. 2018) (fuel efficiency standards); Environmental Defense Fund v. EPA, No. 18-1190 (D.C. Cir., July 18, 2018); Natural Resources Defense Council, Inc. v. Perry, No. 17-cv-03404-VC (N.D. Cal., Feb. 15, 2018); Coral Davenport and Lisa Friedman, "In Rush to Kill Obama Rules, Pruitt Puts His Agenda at Risk," *New York Times,* April 7, 2018, A16. Juliet Eilperin, "Obama's Former EPA Chief Takes Helm of Environmental Group that's Sued Trump Nearly 100 Times," *Washington Post,* November 5, 2019.

14. U.S. Environmental Protection Agency, "Fact Sheet: The Clean Power Plan by the Numbers," August 2015, https://archive.epa.gov/epa/cleanpowerplan/fact-sheet-clean-power-plan-numbers.html#print.

15. Jesse McKinley and Brad Plumer, "New York to Approve One of the World's Most Ambitious Climate Plans," *New York Times,* June 18, 2019.

16. Barbara Finamore, *Will China Save the Planet?* (Cambridge: Polity Press, 2018); Jeffrey Ball, Dan Reicher, Xiaojing Sun, and Caitlin Pollock, "The New Solar System: China's Evolving Solar Industry and Its Implications for Competitive Solar in the United States and the World" (Stanford Steyer-Taylor Center for Energy Policy and Finance, March 2017).

17. Juliet Eilperin and Brady Dennis, "Major Automakers Strike Climate Deal with California, Rebuffing Trump on Proposed Mileage Freeze," *Washington Post,* July 25, 2019; Coral Davenport, "Automakers Plan for Their Worse Nightmare: Regulatory Chaos after Trump's Emissions Rollback," *New York Times,* April 20, 2019.

18. Clifford Krauss, "Trump's Methane Rule Rollback Divides Oil and Gas Industry," *New York Times,* August 29, 2019.

Acknowledgments

This book focuses on the significant contributions of six of the Carbon Dioxide Warriors whose work resulted in the historic Supreme Court victory in *Massachusetts v. EPA*. Although those six played essential roles in that litigation, they would be the first to acknowledge that they were joined in their efforts by dozens of other dedicated and talented attorneys working for states, cities, environmental organizations, public health organizations, and other public-spirited entities. Two important examples are Marc Melnick and Nicholas Stern of the California Attorney General's Office, who devoted as much time and skill to the litigation as any member of the petitioners' legal team. Both are outstanding lawyers who played key roles in the drafting of significant portions of the D.C. Circuit briefs and skillfully served important advisory roles throughout the Supreme Court litigation. The quality of the petitioners' work product was significantly better as a result of their contributions. Massachusetts's Jim Milkey also enjoyed the assistance of two talented attorneys in his office—Bill Pardee and Carol Iancu—whose enormous contributions to Milkey and the overall litigation were invaluable in the drafting of the written briefs and in preparing Milkey for oral argument. Scott Pasternack with the City of New York also played an important advisory role in the Supreme Court litigation, as did Jim Tripp of the Environmental Defense Fund.

The essential role subsequently played by career lawyers at both the EPA and the Justice Department also warrants distinct acknowledgment. These attorneys served as opposing counsel in *Massachusetts,* and they later applied their enormous skill to implementing the ruling in the aftermath of the Supreme Court's decision. They are the backbone of the nation's government.

As the book describes, collaboration among so many attorneys in a case as challenging as *Massachusetts* can come at a significant cost. Conflict within a legal team of many will frequently arise when the odds of winning are low, when it is not clear which is the best of several possible strategies, and when there is good-faith disagreement about which of several talented counsel should have the lead in drafting a brief or presenting an oral argument.

Such disagreements can be especially intense when the stakes are high. The stakes in *Massachusetts* were enormous in terms of both the importance of the underlying issue and the risks posed by an adverse Supreme Court ruling. There is no more challenging or pressing environmental problem than climate change. A potentially devastating loss in the Supreme Court could have significantly set back all climate change litigation and regulation. Human emotions inevitably run high in such circumstances, as they did in *Massachusetts.* That is why the head of a national environmental organization ominously warned Milkey that "the future of the environmental movement is on your head" when Milkey bucked almost everyone and insisted on seeking further court review after the *Massachusetts* petitioners' initial loss in the D.C. Circuit.

Fortunately, the resulting personal acrimony ultimately did not undermine the petitioners' ability to file an outstanding Supreme Court brief and to present a no less outstanding oral argument, and to win the most important environmental law case yet decided by the Court. Everyone involved played a role in that success; they deserve the nation's gratitude.

Many made possible this book's telling of the story of *Massachusetts v. EPA.* I interviewed on the record, on background, or off the record most of those involved in the case, beginning in the 1990s, even before the litigation itself was contemplated. Individuals interviewed included attorneys from the opposing parties, former career employees and political appointees of the EPA and other executive branch offices and agencies, cabinet officials, court personnel, judges and Justices and, only with the judges' or Justices' permission, their staff. Many of those involved in the case generously provided me with thousands of pages of background documents, including email correspondence, handwritten and typed notes of meetings, draft briefs, and other decision-making and litigation-related materials in hard copy and electronic format.

Because many of the individuals with whom I spoke or who otherwise provided assistance did so on background, I can acknowledge here by name only those who agreed to speak with me on the record. The list of those who agreed to be interviewed on the record who warrant my special gratitude must begin with Supreme Court Justice John Paul Stevens and U.S. Court of Appeals for the D.C. Circuit Judge David Tatel—two spectacular jurists. I had been scheduled for an additional interview with Justice Stevens, to whom this book is dedicated, less than two weeks after he passed away on July 16, 2019. For on-the-record interviews, I am also grateful to former EPA administrators Christine Todd Whitman and Carol Browner, former acting EPA administrator Marianne Horinko, former EPA deputy administrator and assistant administrator for air and radiation Jeffrey Holmstead, former EPA general counsel Ann Klee, former EPA associate assistant administrator for climate and senior counsel Joe Goffman, former secretary of Treasury Paul O'Neill, former solicitor general and deputy solicitor general Gregory Garre, and the Carbon Dioxide Warriors—Joe Mendelson, Jim Milkey, David Doniger, Lisa Heinzerling, David Bookbinder, and Howard Fox. Further thanks are owed to the Clerk's Office of the D.C. Circuit and Public Information, Curator's, and Marshal's Offices of the Supreme Court, whose personnel answered my many detailed questions about the administrative operations and history of their extraordinary institutions. Their administration of justice is exemplary.

Many law students provided terrific assistance in the preparation of this book. Harvard Law School students Katie McCarthy, Tsuki Hoshijima, John Greil, Michelle Melton, Annie Madding, Dennis Howe, Ted Yale, Daniel Mach, and Frank Sturges all provided outstanding background research on several topics over several years. Initial drafts of the manuscript benefited greatly from exacting review and extensive and engaging editorial suggestions from Harvard Law School students Natalie Salmanowitz, Jacque Sahlberg, Harry Larson, Lolita De Palma, Julia Jonas-Day, Bess Carter, Chaz Kelsh, and Michelle Melton. Outside readers Jonathan Cannon, Jonathan Dorfman, Ned Friedman, Dan Reicher, and Don Scherer also generously devoted their time and talents to review an early draft of the manuscript, and each offered a host of valuable suggestions that prompted many changes in the structure and content of the manuscript. This book is far better as a result of their scrutiny.

I would like to express my gratitude to my good friend and faculty colleague Jody Freeman for the Harvard Environmental & Energy Law Program's support of my research for the book. Thanks are also owed to the archivist of the George W. Bush Presidential Library and to the staff at Harvard Law School library, especially Meg Kribble and Catherine Biondo, for their valuable research assistance. I am also grateful to Martha Minow and John Manning, former and current deans of Harvard Law School, for summer research grants and a research leave used to research and write the book. During the 2018–2019 academic year, I benefited from the quiet essential to research and writing, first as a member of Jesus College and as a Herbert Smith Freehills Visitor at the Department of Law at the University of Cambridge, and second as an Academic Writing Residency Fellow at the Rockefeller Center in Bellagio, Italy.

Melinda Eakin, my faculty assistant, is without peer as an editor. No transgression, small or large, misses her eagle eye. I am fortunate to have the advantage of her remarkable talents.

Finally, I would like to acknowledge the invaluable assistance of my book agent, Katherine Flynn of Kneerim & Williams, and my fabulous Harvard University Press editors, Thomas LeBien and Joy de Menil. Katherine provided indispensable instruction on how to navigate the unique pathways for publishing a book intended for a popular audience. Thomas and Joy brought the book to closure—Thomas in guiding my completion of the initial manuscript and Joy in wisely persuading me that I was not yet done when I very much hoped to be. Each carefully reviewed every word, sentence, paragraph, and page of my draft chapters to ensure that the story told would be as good as it possibly could be. I am indebted to both Joy and Thomas for their great skills, close attention, and dedication to this book, as well as their good cheer. Louise Robbins of Harvard University Press and the team at Westchester Publishing Services, especially Sherry Gerstein, provided excellent editing to ferret out errors in the manuscript.

Index